BUILDING
HOME

The 5 step journey to building your best lifestyle

Natalie Stevens

First published in 2017 by Grammar Factory Pty Ltd.

National Library of Australia Cataloguing-in-Publication entry:

Creator: Stevens, Natalie, author.
Title: Building home : The 5 step journey to building your best lifestyle / Natalie Stevens.
ISBN: 9780995445338 (paperback)
Subjects: House construction--Australia--Popular works.
            Dwellings--Australia--Design and construction.
            Housing--Australia--Popular works.
            Dwellings--Remodeling--Australia--Popular works.
            House buying--Australia--Popular works.

Printed in Australia by McPhersons Printing Pty Ltd
Book production by Grammar Factory
Cover design by Designerbility
Editing by Grammar Factory
Typeset in 10.5/17 pt Source Sans Pro by Independent Ink

Disclaimer

For my dad, who;

Brings infinite inspiration
Shares endless knowledge
Gives unconditional love

Thank you, I love you.

Contents

Preface

2003 was a great year. It's the year I became a mum. By 2009, my husband Sam and I had brought four beautiful souls into the world. These seven years not only gave me the extraordinary gift of motherhood, but it also gave me the chance to spend quality time with amazing families from a variety of backgrounds. During this time, I came to learn that most families shared the frustration of needing to upgrade their home during a time in their life when they could afford it the least. I could see how families were desperate for a solution that would give them more lifestyle for less. The flame that sparked my passion was ignited the moment I realised that I knew exactly how they could have that.

The philosophy that I bring to you in this book is that the happiest of families are not the ones who have the best of everything, but they are the ones who make the best of everything they have.

My book supports this philosophy because not only does building give us the chance to create instant equity, but it also lets us spend our budget on the areas of our home that will bring us the most happiness. The truth is, building unleashes a mountain of value that most people don't even know exists.

I'm here because I love to show families how building a new home can let them live the very best lifestyle that their finances, environment and personal circumstances can afford. My purpose is to help families *build* their best lifestyle.

And right now, I'm thrilled to be helping *you* do that. Think of this book as you and me being on a journey together. We have so much to see along the way and I promise that you're going to *love* the destination.

Again, I'm thrilled you're here.

Off we go...

Why we all should be living our best lifestyle

Today I am near the end of a few weeks spent travelling in remote Vietnam with my husband and four children. I am once again hit hard by how fortunate I feel to have been born Australian. It sounds like a cliché, I know, but we really do live in the lucky country where opportunity abounds. I'm often inspired by people who talk about how you can follow your dreams and become anybody you want to be, but when I'm travelling in areas like this I feel a sense of guilt wash over me. It just doesn't feel right to be ambitious here. My hopes and dreams pale into insignificance when I see the reality of the everyday lives of people living in a developing country. But then, when I get home, I feel inspired to make the most of the opportunities I've been given, and the opportunity I cherish most is that of choice.

As Australians we are lucky in that we have the valuable gift of choice. The freedom to choose. As Australians most of us have the opportunity to live a fair lifestyle. Some of us will live a good lifestyle and some of us will live a great lifestyle. What things define a great lifestyle is dependent on our circumstances, but what's important is that we are living the absolute best lifestyle that our finances, environment and personal circumstances can afford. It's about being the best we can be. Living our best life. And where do we live most of our lives? In our homes, of course.

But having a home that supports your best lifestyle is easier said than done. We all get distracted by the hustle and bustle of life. Once we have children, we are so busy just trying to keep our head above water it seems impossible to stop, regroup and consider what we really need to do to improve our lifestyle. We know why we need to make a change because we feel frustrated and restricted. The cosy house that was once perfect for mum, dad and bubs is cosy no more. The maintenance of our cute little old house with all the character is chewing up our time and money in maintenance and repairs, and causing us stress because it's yet another item on our long list of things to do. We'd love to have the space to have friends over but with everyone having growing families of their own, this place just doesn't cut it anymore.

With so much on our plate we look for a quick fix. We still want it all to be perfect, but who's got the time to make it happen? The simple solution is to just start looking to buy a bigger place. An easy option. The trouble is, all the places you fall in love with are either too expensive, missing something you need or are just not quite right. Sure, eventually you will settle for something. Will it be perfect? Probably not. And this is the very place most people end up. They settle for a home despite it not being perfect.

The truth is that you will never *find* the perfect home but you can *create* the perfect home. Just like you don't *find* the perfect lifestyle, you *create* it. We might not always be conscious of it, but every decision we make shapes the life we find ourselves living.

Why I do what I do

I believe there is great power between liveability and lifestyle. Function has always played a big role in my life. For example, having come from a large family and now having a large family of my own, I know how a space determines liveability and how liveability affects lifestyle.

When I refer to liveability, I am referring to the home's ability to enable us to live well or, as I often say, its ability to help us live our 'best lifestyle'. If the floorplan has been designed in a way that enables our home to function optimally for us, then I'd consider that home to have good liveability. For example, we enjoy the hustle and bustle of a busy family and it makes sense for us to have an open-living floorplan. But it's very common for parents to want their own retreat or a separate area of the home for the kids to go to. The perfect liveability is unique to each of us. Another example of how space determines liveability is the ever-popular alfresco dining area. Eating outside on a beautiful summer evening is an unsurpassed joy for me and it makes a real difference to how we spend time as a family. Our lifestyle wouldn't be as good without that space.

Value is also important. When the same product is delivered more cost effectively it creates better value. Building a new home enables us to create space and lifestyle while also delivering value. But value on its own isn't enough. It's having pride in the space we live in that contributes to our happiness.

As someone who loves to help families maximise their new home budget, I am most inspired to explore ways of creating value at the intersection of liveability and lifestyle. It's not as widely known as it should be that you can utilise modern building systems and processes to create liveability for a price that delivers value. You just need to know exactly where to start, when and where to turn, and what to look out for along the way.

Choosing to build unlocks your potential to have the best home that your budget can afford. If you have a large budget, then your options might be plenty. But equally, for those on a tight budget, building makes way for you to have more home for less.

Your family will have a better lifestyle by choosing to build because not only will you have more money to spend on your home, you'll also be able to direct your budget, no matter how big or small, on the areas of your home that will bring you the most happiness. This is why I believe that choosing to build a home – one that will give your family an ideal lifestyle that you won't find in any other home – is the smartest thing that you'll ever do for your family.

But it can also be scary.

Building a home doesn't take a dream, it takes a plan

Listening to families talk about home building for years has enabled me to realise that the apprehension most families feel about building can be narrowed down to five fears.

1. Fear of the time they'll need to dedicate to it.

2. Fear of making the big decisions.

3. Fear of getting ripped off.

4. Fear of what it will cost them.

5. Fear of something going wrong.

It was identifying these five fears that led me to dedicate my work to sharing the process of planning to build a home with families. I saw an industry that was shrouded in mystery and confusion to the average Joe. Based on our own home building successes, we unpackaged the whole process and put it back together again to create a full and remarkable solution that will help families live their best lifestyle. It shows families how to avoid anything that could cause any of one of the five fears. This system is the result of over five years of turning the process inside out. It's called 'Home Build in a Box – the five-step journey to building your best lifestyle'. Our Home Build in a Box comes with a gift that money can't buy. The gift of feeling in control for the entire journey.

This book is going to do two things. It's going to let you in on a few trade secrets about buying and building houses and it's also going to introduce you to the five-step Home Build in a Box system. In PART 1 of the book I'll thoroughly explain the financial advantages of building over buying and also explode some of the myths about the supposed joys of renovating and living in old, character-filled houses. In PART 2 of the book I take you through the Home Build in a Box system, and in PART 3 you'll find templates, checklists and to-do lists that will get you started on your home building journey.

Not where I thought I was going, but where I was meant to be

You may be wondering how I got into the home building business. For the best part of my working life I would have described myself as a potter. All I ever dreamed of being was a potter. What I love most about pottery is that I can start a process with a ball of earth and finish it with a fully functional vessel. What's more, I can do all this relying on little more than my hands and the four elements of water, earth, air and fire. Creating something from nothing, now that's an idea worth pursuing. I'll never forget the day I saw a potter 'potting' for the very first time. It was love at first sight.

But my potting career took a few detours. I've also been an art teacher, graphic designer, sculptor and marketer. Which begs the question: how does an art teacher, potter and graphic designer decide that she wants to show families how they can build a home that will give them more lifestyle for less?

As I mentioned earlier, it all began when my husband Sam and I had our children Archie, George, Matilda and Jimmy. This was when I first began to appreciate that all any family ever wants is to enjoy the best lifestyle that they can afford. What started out as a simple light bulb moment went on to become a vortex of passion and productivity. Suddenly I was immersed in an industry I'd never even considered working in, yet had enjoyed an intimate relationship with my whole life.

You see, what I didn't tell you is that while I was busy making pots, art and babies, my roots were cemented in the civil construction industry. I remember the 'good old days' when I was a little girl and Dad would pile all of us kids into the bucket of the excavator, lift it up and spin it around. He'd go to jail for that today, but can you imagine how much fun that was? Priceless. Growing up in my family meant weekends of driving around the state 'looking at jobs'. In the early days Dad had a plumbing business, sewering towns across regional Victoria. Later, he moved into road contracting and became equipped to produce and construct complete subdivisions. While dad began his civil construction journey in 1967, it's over the past thirty years that he has led our family business in developing approximately 947 acres of residential land, which equates to around 1,040 residential lots. We've also developed commercial properties, such as motels, caravan parks and hospitality venues. Every member of my immediate family works in the family business today in some capacity, including Dad, who still leads the team with his boots on.

Today, as boutique developers specialising in the development of regional residential land, we manage every aspect of the supply chain from the purchase of raw land, through to planning, rezoning, road construction, production and release, and marketing and sales.

Choosing to build a new home in regional Australia enables you to have the very best lifestyle that your budget can afford and that your environment and circumstances can accommodate. In a nutshell, building can give you more bang for your buck.

I believe that as regional Australians we hold the key to a gift not afforded to our city counterparts. The gift of the power to create our own perfect lifestyle. The gift of being able to shape our lives according to our wants, needs and desires. The gift of being able to make our dreams a reality and live our lives not by chance, but with intent.

If you want to feel the magic of walking into a home that holds fresh, untapped energy and that welcomes you with open arms, you should know that building can do that for you. In that very first moment when you step into the space on this earth created especially for you, by you, you'll feel complete. The space will flow, the light will be just right, all the planets will have aligned. Smile, you're home.

Building a new home is just one among many of the journeys you'll travel in life.

Let the journey begin ...

"

The happiest of families are not the
ones who have the best of everything,
but they are the ones who make the
best of everything they have.

PART
1

**Why you should build –
the truth about building**

CHAPTER 1

Building makes financial sense

I was born into the building and construction business and for years I took it for granted that most people knew, that when using the right processes, you can usually build something for less than what it would cost you to buy it. In other words, with the right know-how, building can be cheaper. I know how that might take some of you by surprise. Let's take a look at some examples.

The cost of building

There are three reasons why people choose to build a home:

1. To live in

2. To rent out

3. To sell immediately

Considering the third point for the moment, it's obvious that the main purpose for anyone building a house to sell would be to make a profit. Why else would they do it? If you've ever thought to yourself 'I don't want to be bothered with the hassle of building, but still want something really nice and new', remember this: no-one builds a house to sell to break even or lose money. They do it to *make* money. If you buy a brand new home, it's more than likely that someone is making money from you. If anyone is going to make money out of building a house, why not let it be you?

There are an enormous number of variables that come into play when we look closely at what it costs to build a new home. The most obvious considerations that determine cost are:

- The size, location, quality, limitations and features of your land

- The size of the floorplan

- The materials used to construct the house, including the façade, porticos or pergolas

- The fixtures and fittings within the house

Just to give you an idea of the effect variations can have, I'll share with you a couple of examples from my own recent builds. One is with a volume builder and one is with a custom builder. Both examples are based on present day costs.

Volume builder

The house in this example ticks all the boxes for excellent liveability for a growing family. It's also located in a great family-friendly area dominated by owner-occupiers in an estate with an all-ages playground and wide open spaces. The specs on the house are:

- 25sq
- 4 bedrooms
- 2 living areas
- 2 bathrooms
- double garage
- 700m^2 block of land

Cost of block: $160,000 (including stamp duty)

Base price for house: $218,000 including site costs

Plus $45k of upgrades = $263,000

Plus $10k for basic landscaping and fencing = $273,000

Total cost to build house: $433,000

To buy this home brand new might cost you around $460,000–$480,000 in today's market. Plus, if it's not your first home, you'll pay $19,570–$20,779 in stamp duty on top of the purchase (more about this later). **Total costs to buy house: $480,852–$502,098**

By choosing to build this home a family could save $67,852–$69,098

Custom builder

Let's now consider a house I built in 2016 with a custom builder. It was fairly basic but very nice with a great floorplan and was orientated to capture the light throughout the kitchen and living areas. Like the previous example, it had four bedrooms, two living areas, two bathrooms and a double garage. The land cost $120,000 and the final figures for the house came in at $250,000. We sold the house brand new for $410,000 with the purchaser paying an extra $18,558 in stamp duty. In this instance the purchaser effectively paid $428,558 for a home that cost me $370,000 to build. These facts demonstrate how, when all things are equal, it makes far more financial sense to build rather than buy.

So now that you have an idea of the real cost of building versus buying a similar house, let's look more closely at how you might achieve some of these savings.

Stamp duty secrets

Few people purchasing an existing home are prepared for the shock that comes when they discover how much they have to pay, on top of the purchase price, in stamp duty. Stamp duty is a term used for land transfer duty. Put simply, it's a type of tax that you pay to have a property transferred into your name. The amount you'll pay is calculated as a percentage of the value or purchase price of the property.

Recently, when attending a community forum on strategies for local investment and growth in a regional Victorian town, the topic of stamp duty came up. People were complaining about the cost of stamp duty and how it makes it harder for mum and dad investors to afford property. The conversation went back and forth among the crowd for some time until I put up my hand and softly said, 'Well, if you build a house, stamp duty is not a problem.'

And why isn't it a problem? Because you don't pay stamp duty on your house when you build. You only pay it based on the value of the land, which ultimately saves you thousands of dollars. Yes, I know, it's a shock. I hope you're sitting down.

There was a big look of dismay on the faces of everybody at the forum until the local MP, who was hosting the event, jumped up. I had visions of him giving me a standing ovation for my brilliance in revealing a simple fact, but instead he responded with a speech on how stamp duty is necessary for the economy and how you shouldn't begrudge paying tax and so on. He never once mentioned what I'd said or confirmed whether I was even right. I was made to feel like a dodgy two-bit salesman who was asking the crowd to commit tax evasion. It seems few people really get this whole stamp duty thing and how incredibly simple and *legal* it can be not to have to pay it on your home.

Your personal circumstances affect how much stamp duty you'll pay, such as whether you're buying a house to live in yourself, for investment, as a first home owner or as a pensioner.

Almost every article I read in the media about stamp duty fails to point out that if you build you don't pay stamp duty on the value of the actual house. I have no idea why such a valuable benefit of building new homes is rarely discussed. I can only assume that it is because people just don't know.

The amount of stamp duty you have to pay varies from state to state and in some cases is means-tested according to income and number of dependent children. Let's look at how much the stamp duty would be for an existing $450,000 home, assuming that the purchaser has previously owned a home and plans to live in the one they are purchasing. These figures were correct in February 2017:

- **VIC**: $18,970.00
- **NT**: $20,057.17
- **TAS**: $16,122.50
- **NSW**: $15,740.00
- **QLD**: $7,000.00
- **WA**: $15,390.00
- **ACT**: $11,460.00
- **SA**: $18,830.00

You can learn more about stamp duty in your state by visiting the State Revenue website relevant to your state.

Now let's take a closer look at exactly how much stamp duty you will save by building. The following example is for a home built in Victoria and the figures were calculated in February 2017 using the State Revenue Office stamp duty calculator found at sro.vic.gov.au/land-transfer-duty.

If you purchase an *existing* home for $450,000, the stamp duty payable will be:

- Previous home owner: $18,970

Remember, stamp duty is an *additional* cost, on top of the purchase price.

However, if you purchase a block of land for $150,000 and build a house on it for $300,000 (using the same $450,000 overall spend) the stamp duty payable will be:

- Previous home owner: $3,870 – **giving you a saving of $15,100**

When you build a home you don't pay stamp duty on the house, as no transfer is taking place. You only pay stamp duty on the block of land you buy. And this is where the big saving is. This saving could mean the difference between you enjoying an alfresco dining area, a butler's pantry or a major kitchen upgrade. So you can now see that the money you save on paying stamp duty by choosing to build is considerable.

Grants, schemes and concessions

The Government wants you to build a house. Australia is growing at an exponential rate. According to the co-author of the Coalition's refugee and asylum seeker policy, General Jim Molan, Australia's permanent population is growing by 200,000 people per year. All these people need somewhere to live. This is why the Government offers grants, schemes and concessions to make building a new home attractive to first home owners. These benefits are of enormous help to people wanting to get into home ownership. Let's look briefly at what is currently available to those who build in 2017.

Victoria

- First Home Owners Grant: $20,000 towards building a home or buying a brand new, not previously occupied home in regional Victoria ($10,000 for metropolitan areas).

- First Home Owners will be exempt from having to pay Stamp duty where their property is worth less than $600,000.

New South Wales

- First Home–New Home Scheme: No stamp duty payable on vacant land valued up to $350,000. Concessions available for vacant land valued between $350,000 and $450,000.

- First Home Owners Grant: $10,000 towards building a home or buying a brand new, not previously occupied home.

- New Home Grant Scheme: $5,000 towards buying a new home or building a new home.

ACT

- First Home Owners Grant: $7,000 towards purchasing new and substantially renovated properties or purchases that are subject to an 'off the plan' purchase agreement.

Northern Territory

- First Home Owners Grant: $26,000 towards building a home or buying a brand new, not previously occupied home.

- HomeBuild Access: Subsidised interest rate loan and low deposit loan. Available to finance newly built homes and the purchase of vacant land on which to build a home.

Queensland

- First Home Owners Grant: $20,000 for those building a home or buying a brand new, not previously occupied in home.

South Australia

- First Home Owners Grant: $15,000 towards building a home or buying a brand new, not previously occupied home.

Tasmania

- First Home Owners Grant: $20,000 towards building a home or buying a brand new, not previously occupied home.

Western Australia

- First Home Owners Grant: $15,000 towards building a home or buying a brand new, not previously occupied home.

Keep in mind that this is a guide only and that these benefits change regularly. The information here is shared to illustrate what might be available to you at the time you choose to build. For current and more detailed information including eligibility, value caps and so on, please visit www.firsthome.gov.au.

You can see that the benefits exist to encourage the construction of new properties. If you buy a brand new, not previously occupied home, you can almost guarantee someone is making money from you. There is little reason or incentive for someone to build a home to put straight on the market if not to make money from doing so.

Building makes it possible to create instant equity

So far we've looked at the 'cash-in-hand' advantage of building over buying, but building may also improve your financial situation in terms of your assets and your net worth. I'm talking here about home equity. The equity in your home is the current fair market value of your home minus any mortgage you have owing on the home. Obviously, as you pay down your mortgage the amount of equity you have increases. The equity also increases as and when the market value of your home increases. So, if you borrow money and buy an existing home, you will gradually build up some equity as you pay off your mortgage, bit by bit, and patiently wait for the market to go up. However, if you build a new home, you have the opportunity to create a nice bit of equity instantly. For instance, if you buy a block of land and build a house with a total cost of $400,000, upon completion it may be valued at $420,000. This gives you $20,000 in equity. Over the life of your mortgage, your property value increases yet the mortgage value diminishes, meaning you get to build more equity every year.

Let's look at an example of how a first home owner created instant equity in 2016:

A 400m^2 block purchased during the pre-sales phase:	$105,000
Home build with volume builder:	$230,000
Fencing and landscaping:	$10,000
Total cost:	$345,000
Bank valuation six months after build:	$385,000
Equity created:	**$40,000**

Plus, this first home owner received the $10,000 Victorian First Home Owners Grant being offered at the time, putting him $50,000 in front.

Home equity is handy if you ever want to borrow it, but it also makes it much easier for you to secure a loan. Additionally, it can help you with property investment, which can be a financial boon in your retirement. In fact, it's because building has the ability to create instant equity that it is a popular choice among developers and investors, as they can effectively make instant money and claim depreciation to boot. In contrast, investors who purchase an existing property are usually paying full market price.

How to create equity

So how do we go about creating instant equity? Is it guaranteed? No, unfortunately it's not. You can't build just any house and expect to make money. I hear time and time again from people who built a home to put it straight on the market and either couldn't sell it or didn't make any money on it. In the worst-case scenarios people lost money. That's where the false idea that it is more expensive to build comes from.

To create instant equity, you have to know exactly what brings value. Not just to you, but to the majority market. You have a better chance at earning instant equity if you're able to create a home that delivers a great floorplan with excellent liveability. You should consider not only what brings value to you, but what would bring value to the majority of people.

You might recall in the introduction I shared that I'm inspired to explore ways of creating value at the intersection of liveability and lifestyle. And that it's not as widely known as it should be that you can utilise modern building systems and processes to create liveability for a price that delivers value. Well, that's what creating instant equity is all about. It's about being smart with your decisions, your design and your fit-out.

These are the key points to keep in mind when designing and building your home. However, don't think for a second that this means you are building a budget home or a cheap home. You are building a home that will give your family their best lifestyle. And why is that so important when it comes to creating equity? Because if the floorplan and liveability of the home are able to give *you* a great lifestyle, then your home will give someone *else* a great lifestyle. And that's where the value lies. That's exactly what a valuer considers when valuing a property.

You need to ask yourself this: Will expensive fixtures and fittings add value to your lifestyle? It may add value to your sense of pride, luxury and opulence. But your actual day-to-day lifestyle? No, they won't. We are so tempted during the planning stages to add this and that. Some things are just too hard to resist. Adding hard-to-resist luxuries will add cost to your home, but not enough value to create equity.

I have created instant equity in every home I have built. Sometimes a lot, sometimes not so much. I have achieved that in a variety of building scenarios: a four-bedroom family home with a custom builder; a three-bedroom townhouse for a downsizer with a custom builder; and a four-bedroom home with a volume builder. Apart from my own home I have built houses for the sole purpose of selling them. It's simple to build something for less than what it would cost you to buy it if you know the rules and don't let your emotions cloud your decision-making. It sounds deflating, but the more you use your head, rather than your heart, the more equity you'll create.

Of course, only you can find the right balance, and I don't want you to think it's all about sacrifices. I am currently building a home in a beautiful, family-friendly estate opposite a playground. I've made a decision to sacrifice about $10,000 in equity because I want the house to have big street appeal. I know the gorgeous façade won't add $10,000 of value to my property but, for the integrity of the area, it mattered to me. It's a give and take thing. If I decided to upgrade all areas of the home, then I would most certainly lose money on this job. Other upgrades I chose were a second living area, an alfresco area, a small butler's pantry, a study and space for a caravan, boat or trailer. All these things add value. It's the things that will never change, or the blueprint of a home if you like, that add value. We'll look at this in more detail in CHAPTER 6.

If you are building what I call your 'dream' home, then creating instant equity won't be as much of a priority for you, as you may very well prefer to have all the 'bells and whistles' over equity.

Spend your budget on what matters most

To me, the most significant financial advantage of building a home is that it lets you spend your hard-earned cash on the things that matter most to you. Or as you've already heard me say, you can direct your budget on the areas of your home that will bring you the most happiness. For example, if you have kids who surf in summer and play footy in winter, then having an outdoor shower will change your life. In fact, having a larger than average laundry will too!

For me, my life is bliss when I pop myself up onto my north-facing box-window seat. As I sit, I can feel the energy in the rays of the early morning sun. It stills my mind, brings clarity to my thoughts and, like magic, prepares me for the day ahead. When I was planning to build my home, I created a special space for this very purpose and I see it as being something that contributes to my lifestyle. You will not *find* a house that ticks every box for you, but you can *create* it.

Buyer beware

The title of this chapter is 'Building makes financial sense', but while that may be true for the home owner, it's not true for everybody in the real estate game. Let me explain.

One day I was chatting to a friend who was expecting her third child. They'd been looking for a new home and she was tossing up her options. To buy or build? It's never easy finding the perfect home for your family within your budget.

The next time I saw her I was surprised to find that she'd disregarded building. She was talking about not needing the hassle and that if she could just be patient, a new or almost new home that suited them would eventually come on the market. And then she went on to say that she didn't think they'd be able to afford to build anyway.

What had I missed here? How had she come to that conclusion when people have been building things and selling them for more than they cost to build for centuries? Then it dawned on me. As with everyone who decides they need a new home, my friend had just spent a week looking at houses with a real estate agent.

When we are thinking of upgrading our home, the first thing we do is contact a real estate agent, whether that's to appraise our existing home for its value or just to see what's available in the area. Without a doubt they wield some powerful influence when it comes to the way we think and feel about buying a new home.

In regional areas in particular, the chances are that even without knowing it, you look to agents for advice. But what you mightn't know is that agents have a conflict of interest when it comes to selling vacant residential land over existing homes. Why? Because agents are paid on a commission basis to sell property. An existing home could be worth two to three times more than a residential block of land. If an agent can have their client purchase an existing home over a block of land they will, at the very least, more than double and in some cases triple their commission. What incentive does an agent have to promote home building? It's understandable that the resources of the real estate sales team are directed towards clients purchasing existing homes.

An agent will sell land to someone who is already *aware* of the advantages of building and walks in the door to buy a block of land. Easiest sale ever … done. However, if someone walks in who needs to upgrade and who is exploring their options for available properties, are they going to suggest they build? Most likely not, because that suggestion could see them miss out on a much larger commission. What they are more likely to say to someone who comes in voicing their indecision between buying and building is this: 'I've got a property that's almost brand new, in the area you're looking at and you can save yourself the hassle of building.'

Result? Tick, commission more than doubled. You can't begrudge them for that; it's their livelihood and any smart businessperson will work to sell their highest return product.

When you meet with a real estate agent, they'll ask you lots of questions to find out exactly what you are looking for. But here are five questions they won't ask you.

1. Do you want a house that's been tailor-made for you?

2. Would you like to save anywhere up to $20,000 in stamp duty?

3. Do you want the opportunity to create instant equity?

4. Do you want to spend your budget on the areas of your home that will bring you the most happiness?

5. Do you want a property that will be maintenance free for five to ten years?

If you answered 'yes' to any of these questions, then building is for you.

Of course, many of you may still go and buy an existing home. But I am hopeful that after reading this book, you'll make an informed decision based on real facts.

CHAPTER 2

Building makes life easier

Okay, so far I've helped you understand how building a new home makes a lot more financial sense than buying a similar one. But money doesn't buy happiness, right? Exactly. Well, I'm here to tell you that building will not only save you money, it will save your sanity because it's less stressful than renovating or living in a gorgeous old house that needs constant maintenance. It also helps you get closer to living your best lifestyle because *you* decide where the money in your home should be spent. Let me elaborate.

Building beats renovating

Build new or renovate; that is the question. When you're faced with that question, you may believe that renovating will be less stressful. After all, you've got a house you can live in and won't have to find somewhere to live while the new one is being built. And if you're renovating you'll only have to alter a few things, or maybe a lot of things, depending on what you discover as you go along. You can make do with a plan on the back of an envelope instead of hiring an architect or draftsperson, right? Well, if you think that renovating is easier than building new, let me tell you a tale.

A cautionary tale – falling for the old, character-filled home
My husband Sam and I bought our first home in 2000. It was the most exquisite little church in the heart of town. As on many occasions in my life, it was love at first sight. The energy was incredible and upon inspection we absolutely knew that it would become the place we'd call home. It went to auction and we ended up spending $60,000

over our budget of $100,000. We paid $160,000 for what we thought was the most magical fairytale home in the heart of town, but most people couldn't believe we paid so much for it. Three years and two children later we sold it for $375,000. Turns out it was a good buy after all. But we were also lucky.

Our love affair with old, character-filled homes continued and in 2004 we paid a premium price for what was essentially a redundant shell of a 1920s Edwardian on a 1600m² block in the centre of town with views as far as the eye could see. It was one of the few remaining undeveloped blocks in Warrnambool. All our dreams were about to come true – we were on the cusp of the rebuild of a lifetime on a block that would have our children feel like they lived on a small farm, yet had the convenience of being in the middle of town. We knew that realising our vision wouldn't be simple or easy, but we felt prepared for the challenge. I'd been around home building and renovating my whole life. It was a big project, sure, but how hard could it be?

While in a heritage precinct, the home itself wasn't heritage listed, meaning that we needed to comply with the planning requirements that were related to maintaining the integrity of the streetscape. 'Awesome', we thought; we love the streetscape. But what was about to unfold was beyond our worst fears. When we submitted our first planning application, the heritage advisor contracted by the Council felt that the home should be heritage listed and therefore not touched. Um … but what about the fact that we have it in writing, in black and white, that it's not heritage listed? Hello, um ... excuse me, but it's not heritage listed. 'What's that?' I can still hear them say. 'You'll need to talk to our heritage advisor.' What? They couldn't possibly mean we ask the same person who had just said I don't want you to touch the unliveable, no-bathroom, no-kitchen home that we'd just bought despite all the planning documentation we received from Council prior to purchase to say that we can? Oh dear. Oh crap. Holy cow. We had just paid a fortune for something we couldn't touch. Ouch.

My heart couldn't bring myself to scrap the idea of the renovated farm house: the chooks, the vegie garden – you know the vision ... baking cookies, babe in arms, singing songs from *Calamity Jane*. We could have easily taken the developer's approach and kept the cottage and cut up the backyard like everyone else who had cashed in on the old blocks, but I just couldn't bring myself to abandon the vision I had of my family living there. What transpired was three years of persistent, gut-wrenching determination to make that house our home. We employed our own independent heritage advisor who took on the Council in assessing our application. He became our knight in shining armour as he ensured that its merit was assessed on the facts. It was a devastating experience and while we went on to build a magnificent home that remained sympathetic to the period in which the original home was built, it cost us so much time, money and pain. I was pregnant with our third child and four weeks short of my due date when we finally moved in – *three years* later.

There's no question about how much we love our home despite everything we went through, but it took some time for us to recover from the angst of the whole experience. And what's interesting is that it wasn't building the house that gave us the grief. It was all the baggage that came with renovating. All the planning, approvals, back and forth and back and forth. If I could write a list of things you can do to make your home building journey a nightmare, this is it. Moreover, had I known how that house would create one of the most challenging times we'd ever endured I don't think we would have done it. It was purchased purely on emotion, 100 per cent. Today, looking back, there's no doubt it turned out to be one of the most rewarding things we've ever done, but it was tough. Really tough. If there was ever a hard way to build a house, we did it. I could almost write a book on that experience alone.

The process from purchasing this property to moving in took three years of frustration, despite the fact that I was experienced in building and construction and had all the contacts I needed. And it all happened because we purchased a home that had 'character'. But this process did give me a gift – the gift of comparison. Now I'd built the new and renovated the old. Now I had both sides to the story. It put me in a far better position to be able to be where I am today. I couldn't be here now sharing my thoughts with you had I not had the experiences of building that involved the good, the bad and the ugly.

From that time on, I knew there were three ways to own a home.

- Buy – pay full market value

- Build new – pay what it cost to build

- Renovate – open Pandora's box

So consider my advice: In many cases, building, done properly and with the right support, can be a better choice than renovating.

Maintenance madness: when the love affair loses its spark

The stress I'm talking about in this chapter isn't just related to renovating and building. Remember that all houses require maintenance, and the hassles that you inevitably face in the future when something around the house needs fixing are yet another reason to choose building new over buying an existing, older house.

My good friend Natalie (they call us 'the two Nats') is a case in point. Nat lives in Northern Victoria. Recently when we were on a plane taking our annual getaway together, she asked me a question that made me laugh.

'What is it that you really do, anyway?'

After twenty-five years of friendship she never really knew what I did? Funny, but a part of me liked that.

When I told her that I help families who need to maximise their home budget have a better home, in a better location, for a better price, I was not surprised how she went to town on the challenges she was facing with her existing home on a daily basis. This is what she said:

I love my house but in reality it is just maintenance, maintenance, maintenance. Our garden is great, well it could be, but it's just never going to be because I'll never have the time. Ten years ago when we first bought a house, I wanted a house with character and that's exactly what we got, not knowing what it was going to be doing to me in a few years' time.

Before kids, we had all the time in the world to potter about and get stuff done. Maintaining our home was our hobby. But now, it's different, I just want a simple, low-maintenance garden. I mean, it's not like I'm a stay-at-home mum anymore.

Nowadays we have this massive, amazing, overgrown yard that we're never in. Any time we get to spend in it is working in it, rather than enjoying it. Our home has become one big chore rather than a sanctuary.

Bills, bills, bills. It's not a cost-effective home. Our old appliances suck up a lot of energy. We have trouble with our wiring and the list goes on. We work just to cover bills and have little left over to play with. Nothing about my home feels fair anymore.

I felt shocked. This happy-go-lucky, successful businessperson raising four terrific kids was feeling swamped and overwhelmed by the time, effort and money her house was costing her. How many other Aussie families are experiencing the same frustrations? Not the ones who've recently built a house, that's for sure.

Natalie went on to explain...

> *My current mentality is being at work and always feeling like 'Oh, I've got to go home and do this and do that.' It's mentally draining. And these are often things we can do ourselves! It's the constant 'We need to do this and we need to do that' and the constant thinking about all the things that we have to do. Not to mention what it's always going to cost.*

> *I always think that, when I'm older, wouldn't it be great not living like this.*

One way to avoid ending up like my friend Natalie is by deciding to build a new home, rather than buy an old, existing one. While old homes will require ongoing maintenance, new homes can generally be considered maintenance free for the first five to ten years. And on a related point, if you buy an existing home there are bound to be some immediate changes that need to be made to make the home suitable for you. Maybe it's a pergola over the back door or a wall knocked out here or there. These all add up in terms of both costs and frustrations.

Old homes cost more to run

Old, character-filled homes can cost you a lot in terms of maintenance, but they can also cost more in terms of energy efficiency and send your power bills soaring. The following is a list of just ten ways in which an old home can cost more to run than a new one:

1. Darker rooms in older homes require daytime lighting to supplement the limited natural light.

2. Single-glazed windows let heat escape in winter and coolness escape in summer which increases heating and cooling costs.

3. Poor-quality insulation with low resistance value leads to inefficient heating and cooling.

4. Old appliances use old technology, which makes them less energy efficient and creates higher running costs.

5. Old plumbing fixtures and appliances use excess water and drive up energy expenses. A tap that leaks one drop per second can waste up to 12,000 litres of water a year.

6. Converting an existing home to solar power requires a significant financial outlay.

7. Traditional building materials such as timber have poor thermal retention and require greater energy consumption for heating and cooling.

8. Poorly sealed windows, doors, vents and floorboards can lead to a fifteen to twenty-five per cent heat loss in winter.

9. Older homes often have European or English gardens that are high maintenance and require high water consumption.

10. Maintaining and repairing older homes takes time. What is your time worth?

Building lets you have what you want

We all dream about living the good life. In the perfect home. But what does that really mean to you? If you could build your dream home, and money was no object, what would that perfect home look like? Would it be bigger? Have more bedrooms? A huge backyard for the kids? Is location important? Do you need room to grow your family? Space to entertain? Or do you simply dream of a clutter-free life in a modern, low-maintenance space?

Choosing to build means ticking as many of the boxes to having the perfect home as your budget can afford. In 2016 we conducted a simple survey asking regional families what would make a perfect home for them and what effect that would have on their life.

Everyone answered the same two questions, filling in the blanks in these two sentences:

1. If I could have the perfect home today, it would have

   _____.

2. If I had these things, my life would be _____.

The overwhelming majority of families wanted more space. That can present itself in various ways, from an extra bedroom or two, an ensuite, a study, a garage with room for more than one car *and* a workshop, or just a larger backyard for the kids to play in or to entertain friends and extended family.

Also high on their lists was the desire to escape from years of paying rent on someone else's mortgage, endless rounds of house inspections by landlords and agents, and the constant need to move on and pack those boxes over and over again. Getting trapped in the rent cycle is a frustration we can all do without.

Others wanted to have a choice of neighbourhood; a few even had a specific street in mind. Many families dream of peace and serenity, friendly neighbours and views worth watching. Their perfect home is a place where they would feel safe and secure, and be somewhere to relax and unwind from the stresses of daily life.

Another dream was affordability, not only in the cost of the home itself, but to have a more energy-efficient home that used the latest technology.

But one thing is certain – everybody agreed that if they had these things, their lives would be blissful, awesome, more comfortable, healthier, less stressful, more satisfying. For many of them, the word was 'complete'.

When I asked my friend Nat what her life would be like if she lived in a brand new home that provided space for her family to grow, with a low-maintenance yard that they could enjoy at their leisure and all within a budget that they determined, her answer was simple.

> *I would be able to do more. It would be mentally different. My life would feel stress free from not having the burden of the endless list of crappy jobs that need to be done on my home. I would be able to bring so much more time to our family life and mentally I would be a different person…a different wife and a different mother. I would feel in control.*

I advocate for building because I know that it presents a great opportunity for families to have more lifestyle for less. And, I trust that by now you're well on your way to understanding why it's a great idea to build your next home. Now I'm going to share with you how to make it happen.

66

Building a home doesn't take a dream, it takes a plan. This is the plan that helps you live the dream.

The Home Build in a Box system – the five-step journey to building your best lifestyle

Home Build in a Box

Now that you know how building your next home can be a great option, I'm going to introduce you to my five-step system for making it happen simply and efficiently. My five-step methodology guides you through the process of planning to build a home that will help you live a better lifestyle.

When using this simple and straightforward system you won't feel the five fears of building, but instead you'll feel in control throughout the entire planning process. Ultimately, it will make it possible for you to benefit from the financial advantages of building and help you create a home that is perfect for you and your family.

Here I will explain the methodology for each step. In turn, each step will direct you to the relevant pages in PART 3, where you'll find valuable resources and straightforward instructions that will make planning to build your home a breeze.

CHAPTER 3

Step 1 – Budget

The reason why some families procrastinate with building is because they don't know where to start. People wrongly assume that the first thing you should do when thinking of building is to engage a builder. In fact, that is the last step in our five-part plan. There are four critical steps you need to complete prior to getting serious about choosing a builder. Taking these steps in the order set out will guarantee to save you time, money and stress.

Your journey to having a house that will deliver your best lifestyle starts with your budget. Your TODAY budget. It is what it is and you need to get it right so that you can have a realistic expectation of what the very best lifestyle for your family looks like today. Not tomorrow, but today.

In our dreams our home is built according to our desires, but in reality, our home is built according to our budget.

Don't pack more than you can carry

So how much is your budget? If you're an existing home owner, you'll have the sale price of your current home to consider. However, as a first home owner your budget is a simple equation:

$$\text{Savings} + \text{grants} + \text{loans} = \text{budget}$$

The first two items in that equation are pretty obvious. You know how much money you have in the bank, and once you've done your homework, you'll know how much you're eligible for in grants and concessions (see CHAPTER 1). The third part is a little cloudier. It all depends on how much the bank, or another person or institution, is willing to lend you. And you should also factor into that how much you feel you can manage in repayments.

Before you even begin to look at what you can build, think about what you can reasonably afford. Write out a list of your monthly expenses. The lender you choose will ask you for this anyway. They will have their own way of calculating your average monthly spending habits, but you may not be typical. Most importantly, be honest with yourself. When you are deciding how much you can afford, think about what might happen if your circumstances suddenly changed – say you lost your job, or fall ill. As a general rule of thumb it is recommended that your monthly repayments be less than a third of your monthly disposable income.

Let's quickly look at some hard facts.

A home loan has two parts:

The principal – the amount you borrow to buy your property, and

The interest – what you're charged by the lender for borrowing the principal.

In some circumstances, such as for investment purposes, it is possible to pay only the interest component on your loan. However, given that the intention of most home owners is to pay off their mortgage over the course of their lifetime, home owners generally pay both the interest and a percentage of the principal with each monthly repayment.

Interest rates have a significant impact on your repayments and it's important to understand that they can vary. Since 1990 the cash rate has been set by the Reserve Bank of Australia from anywhere between 1.5 per cent to as high as 17.5 per cent, with an average of about 4.1 per cent over the past twenty-six years. And while interest rates haven't been above ten per cent since 1991, it's still not a bad idea to have a look at what that kind of scenario might mean for you.

Let's now look at an example of a loan for a mid-range four-bedroom, two living, two bathroom, double garage home on a 500m²–700m² block in regional Australia built to a $420,000 budget.

With an **$84,000** deposit, it would be necessary to borrow **$336,000**. For a thirty-year loan, the repayments would work out like this:

Interest Rate	Monthly interest only repayments	Monthly principal & interest repayments
3%	$840.00	$1,416.59
4%	$1,120.00	$1,604.11
5%	$1,400.00	$1,803.73
6%	$1,680.00	$2,014.49
7%	$1,960.00	$2,235.41
8%	$2,240.00	$2,465.46

Interest Rate	Monthly interest only repayments	Monthly principal & interest repayments
9%	$2,520.00	$2,703.53
10%	$2,800.00	$2,948.63

ASIC's moneysmart.gov.au is a terrific resource that can help you establish how much you might be able to borrow. You can jump online to do some basic calculations.

All lenders have guidelines that affect whether your loan application is approved and how much they will lend. Some are more flexible in their approach than others. All lenders will want to know a few basic facts about you, including whether you:

- Have stable, and preferably full-time employment

- Have 'genuine' savings

- Can afford the repayments on the amount you wish to borrow

- Have building plans that make for a sound investment

- Have a good credit rating

- Exhibit any 'risky' behaviour – other loan application rejections, job hopping, gambling, to name a few

The next step is to decide who will lend money to you. When I ask families who they borrowed from and why, I'll often hear them say,

'Oh, we already had our banking with them and it was easier' or 'That's where we have our existing home loan so we are just going to stay.' But lending is big business and lenders want your interest repayments in their pocket. Why not consider doing your homework and starting again? Find out who will offer you the best deal, not only in terms of interest rates, but also conditions. Make sure you're looking at comparison rates, which are a true indicator of what your interest rate really is, as they incorporate fees and hidden costs.

Types of lenders

You also have a few options when it comes to the type of lender you'll borrow from. We'll look at the following three here.

- Banks

- Credit unions

- Mortgage brokers

Banks

You may have heard what we commonly refer to as the 'Big Four' banks and it's likely that you already have a transaction account with one of them. But that doesn't mean that you have to take your home loan out with them; you can explore your options. While the Big Four banks offer a wide range of products, such as loans, credit cards and transaction accounts, their interest rates can be less competitive due to their popularity. Banks operate for profit.

Banks offer the following advantages:

- They are well-established and hold the largest share of the home loan market.

- You may already be a customer of one of these institutions and this offers easier access.

- The Australian Prudential Regulatory Association (APRA) holds these institutions accountable for all financial promises they make.

Credit unions

Unlike banks, credit unions are not-for-profit organisations. They are run specifically to benefit their members. You are generally required to become a member when you take out a loan, sometimes for a small fee, which can be as little as $10.00. Because they are run to benefit members, they can usually offer more competitive interest rates, lower fees and more personal customer service. Like banks, they are also governed by the APRA code of practice.

Credit union lenders generally offer the following benefits:

- More competitive interest rates

- More flexible lending criteria

- Lower set-up costs and ongoing fees and charges

- A strong focus on personal customer service and more flexibility to tailor the loan to your particular circumstances

- The ability to assist you even if you have had a loan application rejected in the past

Mortgage brokers

A mortgage broker seeks, finds and compares home loan products that will be best suited to your needs. They are the go-between agent between the borrower and the lender.

Mortgage brokers offer the following advantages:

- They may save you time by doing the legwork for you.

- They may have access to rates and terms you might not have known about.

- They negotiate the rate and terms on your behalf.

At the end of the day it's all about finding the best fit for you. I'm certainly not in the business of giving financial advice, not by any means. But I do want you to know that you have options.

Getting pre-approval for your loan

One of the first things you'll need to do is get pre-approval. That is, you need to know what the bank will lend you. Your pre-approval amount sets your budget.

The amount you'll be able to borrow is calculated on your personal financial position across the board. Applying to borrow money is an information-gathering exercise. It can also be a time-consuming exercise, and you can waste time if you have to have numerous meetings with the bank because you don't have everything they need to assess your situation. This time wasting only contributes to the 'hassle' of building and I want to help you to avoid that. Having all the right information on hand from the start saves you loads of time.

Based on the information you give them, each lender will advise what they can offer you in pre-approved funds. You choose the option that best suits your circumstances. Not only should you be interested in the best rates and the best service, but your loan must have the level of flexibility that you desire, such as the ability to redraw on the loan if you wish or to make bulk repayments. Are they offering you the option to borrow with a construction loan (more about this later)? If so, how will that be structured? Be sure you know what position you'd be in if you had a sudden change of events that affected how you serviced the loan.

To organise pre-approval for your home loan, follow the instructions on page 125 in PART 3. You'll also find a complete list of what you'll need to provide when making your application.

Home loan or construction loan?

When you are building you have options for the type of loan you choose. You can have a standard home loan where you borrow the full amount of the money required to complete the whole project, or you can purchase your block of land using a home loan and get what is called a 'construction loan' to finance the construction of your home. There are significant advantages to using a construction loan.

When you take out a home loan to finance your entire home building project, including the land, you will generally pay both principal and interest on the full amount from the beginning. When you take out a construction loan you are advanced the funds as required for each stage of the build. The beauty of this is that you'll only pay interest on the instalments that have already been advanced to you. This means you will save on interest repayments during construction. You'll only be paying interest on the full amount at the end of the build, by which stage the home loan you took out for your land and the construction loan can be combined into one single home loan.

The chart below makes a comparison between normal home loan repayments and construction loan repayments over an eight-month period. These figures have been taken from a four-bedroom home I built in 2016. This example excludes the land component, as a construction loan is only relevant to the actual building component. This chart is also a good way for you to see the breakdown in costs of building a home and how the payment structure of a custom built home might look. (Payment structures and breakdowns vary from builder to builder.)

Month	Instalments advanced on construction loan	Total amount borrowed	Monthly Construction loan repayments @ 4.5%	Monthly repayments on regular home loan of $263,645 @ 4.5% over 25 years	Saving
Plans & building permit	Month 1	$4,120.00	$4,120.00	$33.00	$1,475.00	
Deposit to builder	Month 2	$10,417.00	$14,537.00	$91.00	$1,475.00	
Base stage Builder	Month 3	18,524.00	$33,016.00	$194.00	$1,475.00	
Frame builder	Month 4	$31,251.00	$64,312.00		$1,475.00	
Extra concrete	Month 4	$18,113.60	$82,425.00	$468.00	$1,475.00	
Lock-up builder	Month 5	$72,920.00	$155,345.00	$873.00	$1,475.00	
Variations builder	Month 6	$9,699.00	$165,044	$927.00	$1,475.00	
Fixing builder	Month 7	$52,086.00	$217,130.00			
Completion builder	Month 7	$20,260.00	$237,390.00	$1,329.00	$1,475.00	
Flooring & blinds	Month 8	$10,516.00	$247,906.00			
Fencing	Month 8	$4,103.00	$252,009.00			
Appliances	Month 8	$4,636.00	$256,645.00			
Landscaping	Month 8	$7,000.00	$263,645.00	$1,475.00	$1,475.00	
Total repayments over 8 months				$5,354.00 over 8 months	$11,800 over 8 months	$6,446.00 over 8 months

Note how the loan repayments for borrowing $263,645 upfront are the same each month and are initially much higher than the monthly repayments for a construction loan in which small amounts are advanced each month. Over eight months, the total savings in the construction loan option are a significant $6,446.00.

It's worth noting at this point that while many land owners and builders join forces to market their products as 'house and land packages', they aren't necessarily *legally* bound as a package. In many cases land is sold separately from the house (despite some marketing implying otherwise), meaning you can cut and paste plans onto any available lot. The ideal scenario is that you'll be buying the block of land separately from the house. In this case you'll have a contract with the land owner to buy the land and a building contract with a builder to build the house. This makes it possible to take out a construction loan for the construction of your home.

Signing on the dotted line

When you find a block that is perfect for you, you can purchase it using an 'unconditional' contract or a contract with a condition, such as 'subject to finance'. When you buy something 'subject to finance' you are not penalised in the event that you are unable to secure your finance within the specified timeframe—usually fourteen days. However, one of the aims of our system is to avoid such heartache. So, if you are in any way unsure about any changes that might occur in your financial position, include the condition 'subject to finance' in the contract. Then take the contract of sale to your chosen home lender and make an application for finance.

When you're ready to start building you'll need to apply for your construction loan. However, you can only apply for a construction loan if you have an active contract including a payment schedule with your builder. You must also already have a building permit. In other words, you need to be ready to go. If your circumstances haven't changed and little time has passed since you got pre-approval, you should be able to rely on the information already submitted to your lender when you were establishing your budget. If more than three months have passed, your lender will probably need updated information.

Budget barriers

With any luck, your savings plus your grants plus what you can borrow from a traditional lender will be enough to build the home you want. But what if it doesn't work out that way? Let's now have a look at what to do when the budget won't stretch to what you need, want and desire.

Dream home versus 'now' home

People without any building experience tend to think that building a new home will be a long, difficult and stressful time that they won't want to repeat. However, waiting until you can afford to build your 'dream' home with all the bells and whistles could mean that you spend years missing out on the big advantages that building can bring your family today. Upgrading your lifestyle can be done in increments throughout life. To demonstrate my point, I'm going to let you in on a little trade secret.

The process of building a home is, for many, addictive. The ability to create instant equity with each build, along with often being able to sell a home for more than what it cost you to build it, means that nothing is lost and much is gained. I'm not suggesting that you'll necessarily build a whole bunch of houses, but there's every chance you'll build more than one. Based on a regional survey in the Victorian town of Warrnambool, which has a population of 34,000 and was rated the third most popular sea-change town in Australia by *The New Daily*, ninety per cent of people who had built a home there would do it again and all of them said they'd recommend building to a friend.

The absolute best reason to build a home is that it allows you to create the very best lifestyle for your family that your budget can afford, be that $300,000 or $3 million.

The size of your budget is irrelevant. It's the value-for-dollar spend that is important. Provided that you're sensible in prioritising your wants, needs and desires, you'll get a far better bang for your buck just by choosing to build.

So, rather than wait until you can afford to build your 'ultimate dream home', why not build a home that you can afford now and build the dream home later. Besides, building is always so much more fun the second time around.

The deposit dilemma

Once upon a time, 100 per cent home loans were available in Australia. But these days, while there are ways to borrow 100 per cent of the price of a home, it is not as straightforward as it used to be.

Today, most lenders offer home loans at ninety per cent or even up to ninety-five per cent of the total amount you are looking to borrow. This is called the LVR, or Loan to Value Ratio. The LVR also determines, in most cases, whether you have to pay Lender's Mortgage Insurance (LMI). LMI protects the lender from a default on your part. It does not protect you in any way. So, the mortgage insurer is effectively taking on the risk for the lender.

As a rule, LMI is required when you don't have enough deposit, typically less than twenty per cent. Along with your deposit, most lenders will also expect to see proof of consistent savings for a period of at least six months. If you are required to pay LMI, the mortgage insurer may impose similar conditions.

These conditions push a lot of borrowers, and particularly first home buyers, out of the market.

So, what can you do?

Make sure you've thoroughly researched your Government grants and incentives. In many cases, the First Home Owners Grant is only available to those who build or buy a brand new, never previously occupied home and comes in handy when you are scraping together your deposit.

The best approach, however, may be to turn to your family.

Family favours

The easiest way around the deposit dilemma is to garner the support of your parents or close family members who already have considerable equity in their own homes, *and* are willing to support you by 'guaranteeing' your loan. In other words, you need someone to 'go guarantor' for you. Essentially, a guarantor is a close family relative, usually your parent/s, who has either paid off their own home or has a significant amount of equity in it. They must be willing to put that home up as security against your loan, and agree to cover your loan repayments if you default.

Most lenders offer guarantor loan products for first home buyers, although they go by different names, depending on the financial institution. Terms you might hear that refer to these types of arrangements include:

- Family Pledge

- Family Support

- Family Equity

- Fast Track

- Family Guarantee

Guarantor loans have many benefits for the borrower:

- You can get into the housing market more quickly

- You can borrow the money to cover your deposit if you don't have enough savings

- You can borrow 100 per cent of the cost of the purchase

- You can avoid LMI, which can be significant

The main risk for the borrower is that they may be enabled by the guarantor to take on a larger debt than they can really afford.

The guarantor, however, faces a number of risks:

- They agree to put up all or part of the equity they have in their own home as security for the borrower.

- They may have to cover loan repayments if the borrower's circumstances change.

- They may need the money back for an unseen change in their circumstances before the borrower is able to repay the funds.

While going guarantor can be risky, there are ways to minimise the risk. The amount of equity offered as security doesn't have to be 100 per cent. It can be limited to an amount that equals the deposit required by the bank. This means that you can go ahead and borrow at ninety-five per cent LVR or less, and the guarantor's security arrangement covers the other five per cent or more. Once you have repaid an amount equal to the guarantee, the guarantor can be released from their obligation to cover your payments, pending approval from the lender.

Another downside of applying for a guarantor loan is that the approval process can be quite lengthy. The lender will need to assess the value of the security being offered and the financial status of the guarantor.

Other ways parents can assist

Apart from actually gifting you the money or guaranteeing your home loan with the lender, there is another option whereby parents can avoid risking their own financial situation or their home. It is a formal, managed loan through a bank or lender and it helps to keep everything above board and official.

This new product is called a Parent Assist loan. Basically, it takes the form of a structured loan between you and your parents and can cover an amount up to 100 per cent of the purchase price, depending on what your parents are willing to lend you.

A typical example might be that your parents lend you twenty per cent of the purchase price: equivalent to a deposit. And then, the bank provides the rest. You repay both your parents and the bank with interest. In some instances, your parents might agree to accept a far lower interest rate than the bank.

Not all lenders offer this kind of loan, and the guidelines and eligibility factors may differ from regular home loans.

There are a whole bunch of baby boomers out there who own their own homes outright and are sitting on hundreds of thousands of dollars in equity. Using this scenario, they can make a real difference to their children's lives. So if you are that special, trusted and deserving son or daughter with equally trusting and deserving parents, then you should ask them about this. What have you got to lose?

We call these options 'creative pathways to home ownership' because that's exactly what they are. Being creative in finding solutions to get you benefiting from home ownership sooner.

Building a home is a huge and exciting milestone in your life and one that is going to affect you in many ways for years to come. The plethora of information available can be overwhelming and I cannot stress strongly enough the importance of seeking independent professional financial advice before committing to any type of construction loan or home loan.

Turn to page 124 and complete the to-do list for Step 1 of the Home Build in a Box system.

CHAPTER 4

Step 2 – Priorities

This step is all about setting your priorities and having the secret weapon you need to easily define your needs, wants and desires.

One of the five fears many people have when building is of making the big decisions. Those big decisions are often made up of lots of smaller decisions. When all those little decisions are spinning around in our minds they can lead us to procrastinate ourselves out of our best lifestyle. Fear of making the wrong decision can stop you from making any decisions at all.

However, when we nail down every decision into our very own checklist we can feel safe and warm in the comfort that only three things are relevant when you are looking for the yes, no or maybe answers. These are:

1. Your needs

2. Your wants

3. Your desires

Needs are the things we must have for our lives to function well. For example, my family of six needs four bedrooms, two living areas, two bathrooms and a double garage.

Wants are the things that aren't essential to the function of our home, but make us happy when we have them. For example, I want an alfresco area and a study.

Desires are the things that would be considered luxuries, such as a butler's pantry or a theatre room. These are the kind of things that you might choose to have in your 'dream' home.

We've already established that, in reality, most of the decisions about the type of home you build are determined by your budget. You can only build what you can afford. But, where the magic of building comes into play is how it lets you spend your budget on the areas of your home that will bring you the most happiness. You can make decisions using a give and take system that will, upon completion, paint the picture of what your personal best lifestyle looks like with the budget you have, be it big or small. This chapter is all about figuring out what rooms, features, fixtures and fittings will contribute to that lifestyle. It's about setting out what you need, want and desire *in order of importance*, so you know how far your budget will take you.

Your secret weapon

To help make both the big and small decisions simple and straightforward, we've created your secret weapon to simplicity – a Builder's Brief template. You can waste so much time, and create so much stress and frustration if you worry about whether you've thought of every little thing. Almost everybody forgets something when building their home, and you shouldn't assume your builders and tradies will think of it during the build. If something's not there, they assume it's because you made a decision for it not to be there. Anyone missing ample outdoor taps?

Because building lets us create a home according to our wants, needs and desires, it's only fitting that we choose to use a system that, within a matter of minutes, makes it easy to define those wants, needs and desires and get them prioritised. You simply fill in your brief according to the instructions. And don't over-think it. It's not like it's set in concrete; it's just a great starting point that has a profound effect on the ease and efficiency of your planning.

The Builder's Brief helps you define what you want in your home, but it has another purpose that is just as important. Having a Builder's Brief is gold when it's time to get quotes from builders. Getting quotes is one of the biggest barriers to building. Why? Because in the majority of cases, builders' quotes are not comparing apples with apples. Quotes from different builders can vary so much that you are left feeling confused and scared that you are being ripped off. It's enough to put you off so much that you abandon your home building journey altogether. No-one wants to get ripped off, right? But with a $60,000 difference between quotes, how do you know who to trust? And how does that discrepancy occur? Let me explain.

We often make the mistake of handing our plans to, say, three builders accompanied only by a verbal overview of what we love and what we don't. We might think that we've been really clear about what we want in our home, but a casual verbal briefing like this can mean different things to different builders. In much the same way that three artists will paint three different interpretations of a landscape, three different builders will give you three different interpretations of your new home despite the floorplan being the same. You might then find yourself throwing your hands in the air because you have no idea where the extra $40,000 here and $60,000 there have come from. On the surface the builders' quotes look the same, except for one thing – the price. But it's very easy to clock up $60,000 worth of extras in a home. The types of windows alone can clock up that kind of money.

Here's where the Builder's Brief makes magic. When you give three custom builders the same three things – your budget, your Builder's Brief and your plans – you'll be presented with each builder's interpretation of what they can deliver for your budget. The magic in this process is that each builder knows exactly what you have to spend, what things matter to you most and the floorplan you want. Each builder then works separately to put together their own version of your best lifestyle. The difference is that *you've* set the scene for the quote, not them. What a beautifully simple way to compare apples with apples. You can learn more about how your Builder's Brief helps you choose the perfect builder in Step 5 of our Build in a Box system.

What if I'm using a volume builder?

If you are using a volume builder, the principles are no different and your Builder's Brief is just as valuable. Volume builders have systemised the whole process and they will guide you through the selection of every possible option. But what's great about having a Builder's Brief, even if you are using a volume builder, is it lets you feel in control from the very beginning. You won't be going in there unprepared for all the options that might be thrown at you. When the options are offered, you'll be able to make a speedy decision about whether you want them, because you have your priorities set down in black and white in your Builder's Brief. Because we are all about making the journey as efficient and straightforward as possible, we want to save you time. Submitting your budget and Builder's Brief to volume builders before you meet them will make your initial meeting a powerhouse of efficiency.

Completing this step empowers you to say to your builder, 'This is how much I have to spend, I have prioritised my new home according to my needs, wants and desires. With all this information please present me with my best options.' Some budgets won't get past the 'needs' section. Others will have it all. Either way, this is a great time to know where you stand. If you make your budget your first step, and your Builder's Brief your second, you're on the right path. What's more, if things aren't shaping up to be what you'd hoped, you can jump out without having sacrificed too much time and effort.

You'll find a Builder's Brief template
that you can fill in on page 128.

The devil in the detail

There are times in life when the little things can matter the most. Living in your new home is one of those times. There might be a few little luxuries that you believe would make a big difference to your lifestyle. Or others that are merely tempting money guzzlers that suck you in at the upgrades stage but don't bring extra value to your lifestyle. Some upgrades you might not even use (think home intercom system; screaming is so much more fulfilling).

Your Builder's Brief is focussed on the core structure of your home, but to help you make sure that you have absolutely every base covered, you should also complete the more detailed Welcome Home Checklist. The items on the checklist are the nuts and bolts of your proposed home; the little things or the big things, depending how you look at it. Whether what's listed in the checklist is within your budget or not, that's what needs to be determined.

Welcome Home Checklist

We call this our Welcome Home Checklist because the things on this list are the ones that can make you feel either warm and fuzzy or cold and irritable. For example, how many of us have arrived home late at night and had the light sensors kick in, saying 'Welcome home'. Yep, that's nice. In contrast, if you arrive home late, cold and wet and you're in the pitch dark, nothing says welcome home at all except the paranoia about bad guys lurking in the bushes!

When we built our house the cabinet maker was always going to come back to put in one of those great bins that sit inside a cupboard and pop open with the touch of a button. I was excited at the prospect of a lovely mass of compartmented plastic being sprung out to greet me in all its glory. Eight years later we still use the plaster-mixing bucket that the plasterer left at the house when he finished working there. I've never considered buying another bin because he's coming back, right? That's a prime example of a little thing that would have meant a lot. Too small to make a fuss about, yet big enough for me to be mentioning it all these years later. There's a tip there, too. Don't let any tradie go until they are finished. It's only the super sweet, hard-to-find ones who say 'I'll come back and fix that' and actually mean it. Hold off their last $1,000 if you have to.

The Welcome Home Checklist is not designed to point out all the little luxuries you'd like but possibly can't afford. It's for the purpose of forming part of your Builder's Brief and is valuable in painting the overall picture for your builder from the beginning.

The items on the list are just things that spring to mind when I think about what might make the perfect home for the target market I build homes for – families. That's not to say I choose all of them, by the way – far from it. Some contribute to a better lifestyle and others are considered luxuries. For example, someone who feels the cold might love the idea of a heat lamp in the ceiling of their bathroom. But for our family, made up mainly of hot bloods, we'd never consider such a thing. This list is relevant whether you are using a volume builder or a custom builder. If you are using a volume builder, you'll see how they have mastered the system of added upgrades. But once again, because we are about saving you as much time as possible, we'd like you to consider some of these things early in the process so you can make swift, informed decisions when the time comes for you to choose your inclusions with your builder.

**You'll find a Welcome Home Checklist
that you can complete on page 132.**

Don't forget!

As well as the previous checklists, we'd like you to give special consideration to the following areas when preparing the information for your builder and setting your priorities. These are things that are easily overlooked during the planning stages, but may make a big difference to the comfort and liveability of your home.

Telecommunications

Telecommunications includes home telephones, internet, data and cabling. Thinking about your telecommunications early can save you big dollars later. These days, effective telecommunications are essential to our lifestyle; however, they're often overlooked at the vital planning stage. Your telecommunications requirements should be considered in conjunction with your electrical plan. We prefer to use a telecommunications specialist who focuses specifically on the most up-to-date advancements and technology.

'Smart home' features

Smart homes are, for many, the homes of the future. Imagine every element of your home from security to lighting to blinds controlled from an iPad or phone, whether you're at home or on the other side of the world. If you like the thought of being able to manage the functions of your home from afar, you'll love what's currently being offered in the world of smart homes. A smart home specialist will be up to date with all the latest technology and can help you decide if creating a smart home will be right for you.

Heating and cooling

It's easy to assume that heating or cooling your home is straightforward. However, there are variables worth considering. While you can chat with your builder about your options, don't assume they are aware of the latest technologies as, like most things these days, the products and trends are rapidly evolving. If you want to be fully informed about the latest heating and cooling technologies, it's worth spending some time visiting a specialist who is a market leader when it comes to offering the latest products.

**Turn to page 127 and complete the to-do list for
Step 2 of the Home Build in a Box system.**

CHAPTER 5

Step 3 – Land

If you're going to build your home – be it your 'dream' home, 'now' home or something in between – you'll need some land to put it on. Unlike the plans for your house, you can't change your land around once you've bought it, so buying the right block is absolutely vital. We get asked a lot of questions about buying land, and the following is based on what our clients most need to know about the all-important matter of choosing and buying the right block of land.

Location, location, location

As Dorothy in *The Wizard of Oz* said, 'There's no place like home.' It's true; the place you choose to call home needs to be a place that will bring value to not only your pocket, but also your lifestyle and wellbeing. You'll want to feel 'at home' with those around you, living among families who share similar values.

There's no doubt that where you choose to call home will depend on things such as price, proximity to schools, shops and recreation. But what is often overlooked is simply the 'vibe' of a place. The feeling of living in a space created especially for your family is golden. Your best lifestyle starts with your best location and because not all land is created equal, it's important to understand what it is about a location that will help you live your best lifestyle.

As a family building your own home, you will most likely be happier in a location where other homes are also owner occupied. You'll love lifestyle features such as playgrounds, parks, open spaces, bike paths and walking trails. In most regional areas, most estates are just a few minutes away from everything you need. Even those that 'feel' further out are usually only minutes from major shops, schools and recreational facilities. In exchange for the extra few minutes' drive, some estates provide you with an exceptional family-friendly lifestyle, with playgrounds, open public spaces and access to public transport.

Never underestimate the importance of the location where you choose to build your home. Because this is so important, we have created a Land Comparison Chart so that you can easily compare apples with apples. You can find the chart on page 145. My biggest tip would be that if you are building a home for you or your family to live in, consider choosing a quality, family-friendly estate that is predominately owner occupied. In many cases (not all), the landscaping can give you a little hint as to whether the homes in an area are owner occupied or not – home owners tend to invest more in their landscaping than renters.

Be an early bird

There are times when land is offered for sale before it is ready to be built on. This is what is known as a 'pre-sale' or 'off-the-plan' purchase and is made before land titles have been issued or before construction of the subdivision and roads is complete. Developers sometimes go down this path because the costs involved in developing a new estate are substantial. By holding back the development of the roads until a certain percentage of the land has been sold, the developer can realise a more immediate return on their investment. This is quite common, and there are several reasons you may consider making such a purchase that can work in your favour.

Think about how the following advantages could benefit your family:

- You don't have to pay for your block until settlement, giving you extra time to save.

- Because you haven't yet paid for the block, you'll have no holding costs such as interest payments on a loan.

- You can secure the best block with a minimal deposit and pay no more until settlement.

- Sometimes developers offer incentives, such as rebates or special offers on pre-titled blocks.

- Buying during the pre-sales phase could provide an opportunity to negotiate the price, as developers may be motivated to sell during this stage.

Buying off-the-plan does require a level of visualisation. If you are buying land this way, ask for a look at the 'precinct structure plan'. A structure plan takes into account the whole area, not just the estate you are buying in. It will include future lifestyle features, nearby future developments and local infrastructure plans. Viewing the structure plan is a key element of practicing due diligence.

Many of our clients also ask us what the difference between a sales plan and a plan of subdivision is, and it helps to understand this. The sales plan acts as the main marketing brochure for the developer. It is usually specific to the stage of development that is currently on offer. A sales plan shows:

- The stage currently on offer and the lot sizes

- Basic dimensions, such as frontages

- Which lots are still available for purchase

- Contact details for enquiries and where you can get more information

The plan of subdivision provides details of the block sizes, dimensions and any other details such as easements. It is usually drawn up by a surveyor. A plan of subdivision shows:

- All the stages that will be offered for sale across the entire estate

- The sizes and dimensions of every block

- Any existing easements that may be present on every block

Suss out your site

Once you've chosen the general location of your land, you need to look at more site-specific matters. In particular, you should check to see whether the land is affected by covenants or easements and whether it is fully serviced.

A **covenant** is a condition that imposes duties or restrictions on the use of that land, regardless of the owner. Covenants can define building heights, fencing types, or even the materials from which the dwelling is constructed. Many estates have general covenants that apply universally to Residential 1 planning zones; however, there are a number of estates that have significant covenants or restrictions. Make sure you know which ones they are to avoid any nasty surprises when you go to build.

An **easement** is a part of your land that someone else may require access to in the future. They could be local authorities such as the Council or suppliers of electricity, water and sewerage services. While the need to access the easement may be rare, the portion of land in question must be kept clear unless you obtain a 'build over permit'. With today's technology, such as the ability to put cameras up pipes, the fear of easements is diminished somewhat, but is still something you need to be aware of. An easement often runs along a property boundary, allowing your home to be positioned to avoid it. Chat to a land specialist if you are unsure about how an easement could affect you.

Connected services include electricity, water, sewerage and telephone, and will ideally already be connected on the land you are purchasing.

Most residential estates are fully serviced and ready to go. However, in a rural or rural-residential area this may not be the case, and you will be required to pay for connection of these essential services out of your own pocket. If you need to do this, make sure you get quotes from several civil contractors before purchasing the land. Ensure that you ask about service connections early on so you can account for the costs in your budgeting.

Site costs

Site costs are the costs of preparing your block of land for building and other site-dependant costs, such as slab upgrades. Some blocks require more work than others, and how much depends on three main factors.

Soil type

It may seem strange, but the type of soil is important as it determines the type of slab on which you can build your home. Far from being 'just a slab of concrete', the slab is highly engineered and the type of slab you need depends entirely on the environment in which it will sit. Things like rock, clay, sand and soil content can affect the type of slab required, and some are more expensive than others. On all accounts it will be necessary to have the soil tested to check on these factors, which will determine the nature of your concrete slab design.

Site preparation

Some of the site excavation work may have been done prior to the land being offered for sale. But if not, you or your builder can make your own enquiries about what the costs to level the site might be. It may be possible to negotiate these costs with the developer.

Excavation costs can be expensive, but if you get the right advice early on you needn't blow your budget.

Land specialists work directly with site excavation contractors and developers. While rare, in some cases developers will agree to contribute to or even include site levelling in the purchase price. If site costs present a significant hurdle for you it might be a conversation worth having with your land specialist, who deals directly with the developer.

Slope of the land

A sloping block can increase your site costs to varying degrees. Don't be put off from buying a sloping block of land for fear of site costs alone, as there can be some really exciting opportunities presented, with homes oriented towards the light and designed to capture beautiful views. Many two-storey houses can be adapted to make use of the fall of the land to create a unique home. Designing your home with a garage or separate living area on the ground floor could allow you to take full advantage of a sloping block.

Regardless of why you might choose a sloping block, you'll need to consider the following levelling options.

- **Site fill:** This is an area where caution is advised. Filling a sloping block requires a guarantee of stability and the quality of retainment must be exceptional. A filled block often experiences a certain degree of movement if strict and thorough procedures are not observed during the operation. Filling can also increase the site costs when the soil is tested for the slab upgrade. The stability needs to be guaranteed by drilling down into the natural ground level, not just the newly created soil level.

- **Site cut:** The cost of this operation will vary depending on the amount of earth to be removed, as well as the ground quality. Obviously, removing rock is far more expensive than removing soft, fertile soil. You are strongly advised to seek the advice of a civil contractor, who will be able to give you a quote. In some cases, it is possible to negotiate the site levelling costs into the purchase price of the block to avoid budget blow-outs. Some developers are open to negotiation, depending on their motivation to sell.

A land specialist who works directly with site excavation contractors and developers can assist you with all your site levelling concerns.

Keep it legal

Conveyancing is the process of transferring ownership of a parcel of land from one person to another. It is a legal process that is performed by a legal conveyancer. In a property transaction there is a buyer and a seller and both party require the services of a conveyancer.

You will need to choose your legal conveyancer *before* choosing your block. You won't pay for any services until your land has been settled, so there's no reason not to get in early. Once chosen, they will be the ones who take care of your needs, making sure that all is as it should be with the contract. They are kind of like your guardian angel, protecting you during the land purchase process. They'll let you know if there is something in the contract that you might need to give extra consideration to and they'll always be on to anything that doesn't seem quite right. Using a legal conveyancer takes the worry about legalities away from you completely, leaving you to focus on the fun part.

A **Certificate of Title** is a legal document that proves you are the owner of your property. Once you have paid for the land in full at settlement, it will pass into your hands and you will be free to begin the building process. Your little piece of heaven is now finally yours!

Engage an expert

If you agree that building gives you the greatest opportunity to live the best lifestyle that your budget can afford, then seeking advice from a person who specialises in home *building* rather than home *buying* makes a whole lot of sense. Home buying and home building are, after all, two completely different processes. A land specialist is a property consultant who works directly with developers and builders to help clients source the perfect block of land.

There's much to be gained working with someone who understands the opportunities that arise from building, such as stamp duty savings, creating instant equity, little to zero maintenance and, of course, the ability to spend your budget on the areas that matter most.

So, if you want the journey towards your best lifestyle to be smooth sailing, get expert advice from the very beginning. Enquire about buying land from a land specialist who understands every element of the home building process.

Turn to page 141 and complete the to-do list for Step 3 of the Home Build in a Box system.

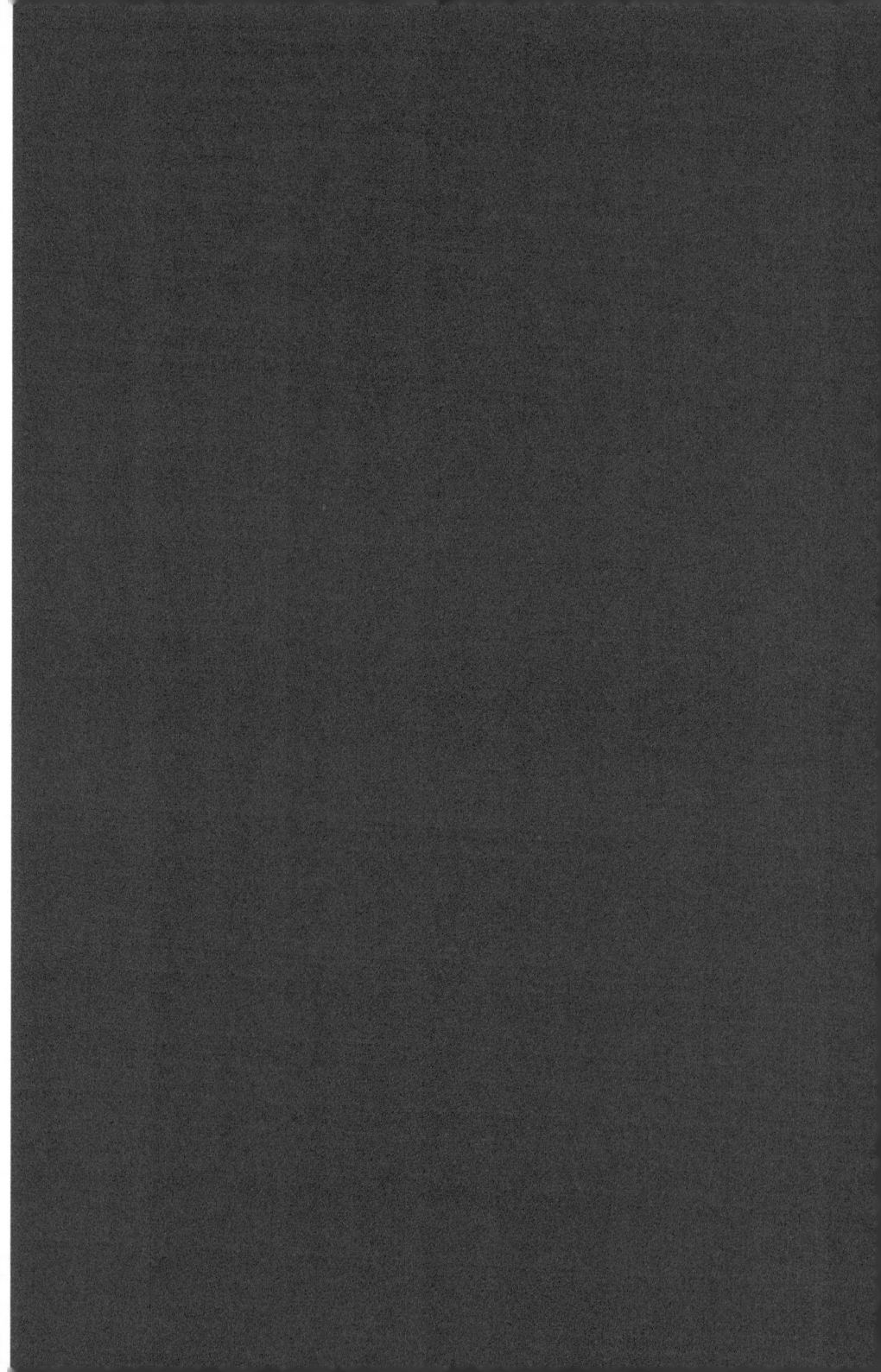

CHAPTER 6

Step 4 – Plans

When building a home, you'll want to ensure it's going to enable you to live your best lifestyle, and your best lifestyle starts with getting the design right. When I talk about design I'm talking about the floorplan and other permanent features of the house. You may have heard about a house having 'good bones'. That is the sort of thing I'm talking about when it comes to design, and getting these features right is the most important step in the five-part system of planning to build your new home.

A great home starts with a great floorplan

Can you imagine walking into home – one that ticks all your boxes and gives your family the space it needs to grow and the freedom it needs to flourish, for a price that delivers value? Choosing to build makes this dream a reality. You might think people fall in love with a home based on the shiny porcelain tiles or the funky feature wall, but the quality of your home isn't just about the tiles, fixtures and fittings. It's true that those things make for a great first impression, but there's one thing more important that a first impression. That's a lasting impression.

A good house is like a good meal. Certain rules apply to make it a meal worthy of enjoyment. Certain flavours go together and others clash. How a home is designed works the same way. There has to be the right flow of space and light. For your home to be perfect, everything needs to flow together in harmony to create the ultimate sanctuary for you. People should always choose a floorplan based on the functionality and liveability it can give their family. A great home needn't be expensive – it only needs to be smart.

My seven top tips for successful design

There are many great books that focus solely on design principles – it's a broad topic. But given that here we are focused on creating excellent liveability for a growing family, I thought I'd share some of my own personal ideas that I think about when benchmarking a home's design.

1. Space

The definition of spacious is not to have *a lot* of space; it's to have *ample* space. Spacious doesn't have to mean big. Any room can be spacious providing it can meet the purpose for which it was created or intended.

How can you make a room appear more spacious?

- Minimise clutter by employing smart storage solutions.

- Use light, low-contrast colours and use a shade of white on the walls and ceiling.

- Avoid drop-down lights. Use downlights instead.

- Use mirrored robe doors. Mirrors bounce light around, making everything appear bigger.

- Have large open entrances into small living areas to make them appear bigger.

- Use cavity sliding doors wherever appropriate.

2. Function

Allow spaces to be what they need to be, not what you wish they could be. For example, many people want a formal lounge or a 'parents' retreat'. But in reality, when we get one of these, we rarely use it. Think about how many family homes you've been in where there's that 'special space'. It often sits there dormant behind the glass doors of what feels like Fort Knox.

As you think about what you need in a floorplan, consider what purpose each space will serve for your family. If what you really need is a rumpus room, call it that and relish in the glory of tripping over brightly coloured plastic toys as you scrunch soft crumbly crayons into the carpet. Trust me, I've been here. To fight it only makes the defeat that much more painful.

Here are some low-value spaces and the high-value spaces that have generally replaced them:

- Goodbye bar and hello open kitchen and alfresco dining.

- Goodbye powder room and hello butler's pantry.

- Goodbye formal lounge room and hello home office with kids' study area.

- Goodbye guest room and hello kids' rumpus room.

As for the theatre room, well, the jury is still out for me on this one.

3. Harmony

Think about the relationship between one space and another. When you are looking at floorplans try to visualise yourself and others using the different spaces at the same time. Your best floorplan should facilitate harmony among the users.

What helps facilitate a harmonious home?

- Separate living areas at opposite ends of the house.

- A family toilet separate from the family bathroom.

- Bedrooms positioned well away from the kitchen and living areas.

- The parents' bedroom close to the second living area so that, in the rare event that you had some time to yourself, you could 'close off' your area.

- Cavity sliding doors that shut off sections of the home.

4. Balance

The walls of a floorplan should align and feel balanced. While homes aren't typically symmetrical, there should be a *feeling* of symmetry throughout the home. No little bits jutting out here and other bits jutting out there. That's just messy and makes the home feel awkward. The more streamlined the design, the more balanced the space.

5. Adaptability

Families grow up FAST. Create spaces that can adapt and change as your family grows. The games room today might become the home gym of tomorrow. Avoid overdoing permanent fixtures that are built into the blueprint of your home, such as built-in entertainment cabinets, bars, bookshelves and desks, to name a few. Keep your options for the future open.

6. Prospect and aspect

Your home should look good from the outside and the outside should look good from the inside. Give consideration to the elements of your façade. Consider the value you place on being 'street proud'. If you're on a tight budget, take onboard one of my favourite sayings – 'less is best, more is a bore' – and jump on the butler's pantry bandwagon instead. (Can you tell I really want one of these?)

As for the aspect, when choosing or designing a floorplan, consider carefully the location of the windows. What will they be looking into/onto? I can't emphasis this enough. You'd be surprised by how often it's overlooked. In fact, this even trips *me* up now and then. Only recently we had to cut down a full-size window and reframe a wall as I had underestimated the overlooking issue and didn't want to have to install unattractive screening over the window. Often it's just a matter of shuffling things around and improvising a bit to get it right. But don't wait until it's too late. Think about this when you are creating or choosing your floorplan.

7. Feeling

When you are designing your home make it your priority to create a 'feeling' rather than a 'look'. The 'feeling' of your home is its innate quality. The thing that makes you smile as you walk in the door each day. It's the part that's solid and unchanging. You create your 'look' later with all your transferable decorating elements, such as furniture and so on.

What things might determine the 'feeling' of a home?

- Paint colours, flooring and curtains

- Ambience of light created by placement of windows and lighting

- Style of fixtures and fittings

- Ceiling heights

- Profile of skirting and architraves

- Openness of space.

- The natural flow of the floorplan

Sustainable design

I don't think any publication written about homes today would be complete if it didn't touch on sustainable living. Many an author has written extensively on the topic of sustainability, but covering that subject fully is outside the scope of this book. However, there are a number of basic energy-efficient sustainable principles that are readily available and can be implemented today. The best time to consider your options is now, in the early planning stages of building your new home.

Here are ten simple ways you can employ sustainable principles in your new home that will reduce your energy consumption and save you money.

1. **Floorplan:** Ensure your floorplan maximises 'daylighting design aspects', which reduces the need for electric light during the day.

2. **Glazing:** Double-glaze all windows to retain heat in winter and keep the heat out in summer for more energy-efficient heating and cooling.

3. **Insulation:** Use quality insulation with high 'resistance to heat' flow (thickness) and ensure it is installed in floor, walls and roof for better heating and cooling energy efficiency.

4. **General appliances:** Ensure all new appliances have high energy-efficiency star ratings.

5. **Water appliances:** Ensure all new water-consuming appliances have a high Water Efficiency Labelling and Standards (WELS) rating. Especially with showers, toilets, washing machines and dishwashers.

6. **Solar:** If suitable, install solar energy when you build, as the costs to do it later are considerable.

7. **Thermal mass:** If you live in a cool area, explore options to maximise external thermal mass for radiating heat throughout the home. If living in a hot climate, consider internal thermal mass that's protected from the sun to keep the inside cool. Thermal mass structures for walls or floors could be natural stone, concrete, brick walls or rammed earth.

8. **Landscaping:** Planting low-maintenance native vegetation that is drought resistant and reduces water requirements.

9. **Rainwater harvesting:** Installing a rain harvesting tank not only saves you water, but also money. In many areas a rainwater harvesting tank is required for a satisfactory energy report.

10. **Greywater recycling:** Greywater is the term used for the recycling of water that has been used in the home. For example, treated greywater that's come from the laundry, showers and taps can be used in the toilet.

For further information on sustainable building practices, view the Australian guide to environmentally sustainable homes at yourhome. gov.au.

A note on orientation

The best orientation for your home is largely dependent on your environment, or more specifically your climate. According to the *Australian Guide for Environmentally Sustainable Homes*, approximately forty per cent of our energy usage is for heating and cooling. The *Guide* also suggests that forty percent could be reduced to zero if new homes utilised sound climate-responsive design.

Given that the Australian Building Codes Board cites eight different climate zones within Australia, let's not delve too far into this here. What's important is that you know that a blanket policy doesn't exist. In cooler climates you'll want to maximise the sun's exposure to warm your home and in hot climates you'll seek to minimise it. Your builder, building designer or architect will be able to help you best orientate your home for maximum energy efficiency.

Choosing your design professional

Now that we've looked at the elements that go into a well-designed home, it's time to consider who will help you design a home that incorporates them. There are a number of different professionals who can design a home that helps you achieve as many as your needs, wants and desires that your budget will allow. Let's look at the three main options.

Using a volume builder

Volume builders offer hundreds of pre-designed floorplans that cater for all budgets. The sales consultant who works with you on behalf of the volume building company will take care of every element, from helping you choose the right floorplan, to plotting the floorplan on your block, to getting all the permits and approvals organised. The volume builder has systemised the whole building process, making it seamless for you as the home owner. It eliminates the need for you to have to 'design' your home. You simply pick the floorplan that will deliver your family its best lifestyle. The work you've already done during the previous stages of the five-step Home Build in a Box system makes this process even easier. Having already defined your budget, wants, needs and desires narrows down your options considerably and you will be offered only the plans that are within your reach, making the process extremely simple and efficient. You won't experience the disappointment felt by others who haven't been through the process, who choose all sorts of things they think they need only to discover they can't afford them. This disappointment can be enough to have families turn their back on building and resign themselves to purchasing an existing home.

Working with a volume builder has many advantages, but you should also be aware of what the limitations can be. To help you decide whether a volume builder is the right choice for you, you should take the following points into consideration:

- You won't pay any additional fee for the plans. The fee is absorbed into the build price.

- Volume builder floorplans are often transferrable from one block of land to another, meaning the same floorplan will work on a variety of available blocks.

- You can visit display homes for some of the most popular floorplans.

- Volume builders are best suited to building on flat blocks that don't require complex site levelling issues. Keep in mind that you can arrange site levelling independent of the builder.

- The copyright of the floorplan is owned by the builder, meaning you can't take their floorplan and have someone else build it.

- It can be costly to modify the floorplan.

- It won't be straightforward to make improvised alterations once building has begun.

Using a building designer

When you work with a building designer you will not be using an existing set of plans, as you would with a volume builder, but working with them to create a unique design. Building designers often work under architects or engineers to transform their ideas into detailed technical drawings, but they can also transform *your* ideas into a complete set of working drawings.

Drafting is the process of creating technical drawings that include the full specifications of the components and elements required to build a home. Drawings will show all dimensions and elevations and include detailed systems of the house such as ducting, plumbing and electrical. A full set of 'working drawings' or 'building plans' make up the document that communicates the requirements of the home to the builders and tradies.

Today, drafting is undertaken using computer-aided design software, making it possible to make changes and alterations instantaneously.

Costs associated with engaging the services of a building designer depend on the scope of the work required. This might include:

- Preliminary sketches

- Detailed sketches

- Alterations

- Planning permit application

- Structural engineering

- Energy rating report

- Structure plans for electrical, plumbing and ducting

- Schedule of construction materials

- Full specification schedule

As a general guide, for the process of going from conceptual sketches to finished working drawings, you could be looking at paying anywhere between $2,500 and $4,000. Costs may vary between metropolitan and regional areas and may take the form of an hourly fee, a fixed price or a percentage of project costs.

You might choose to engage drafting services when you:

- Want to design your home using your own simple ideas

- Have an existing sketch or plan that you want to alter to suit your needs

- Want to plot an existing plan onto your block

- Need to prepare sketches for a planning application

- Need to create working drawings from a plan for a builder

You can find a building designer by visiting the Design Services Association in your state. You will find the websites listed in the resources section of this book. Alternatively, you can explore options in your local area.

Using an architect

Architects are degree-qualified design professionals and generally have the highest level of expertise among your three options. Their professional standards are monitored by the Royal Australian Institute of Architects (RAIA). You can expect them to be the most expensive option, but they are full-service professionals whose fee can cover design, and/or complete project management.

Architects have an expertise that other building design professionals do not, which makes them good choices for complex or unusual designs. You're also likely to get a higher degree of creative input from an architect.

The cost of hiring an architect can vary widely depending on the scope of works required and the level of experience and reputation of the architect. While architects can charge a set fee or an hourly rate, it's common for them to charge a percentage fee. As a rough guide, architects can change between eight and eighteen per cent of the total project fees. The lower percentage might be appropriate when you require more conceptual services such as sketches for planning permit applications, and the higher percentage more appropriate for full design and project management services.

You might want to engage the services of an architect when you:

- Want preliminary sketches to explore the viability of a project

- Want someone who will manage as much or as little of the entire design and construction process as you wish

- Want a more environmentally sustainable, energy-efficient home

- Need a solution to a complex residential need; for example, when building on a steep slope that utilises the fall of the land in the design of the home

- Are building in an estate with a covenant that requires you to build an architecturally designed home that adheres to specific building guidelines

- Want architectural landscape design that's sympathetic to your home design, climate and natural environment

When you work with an architect you can expect that they will:

- Help you set a realistic and viable budget

- Guide you through the town planning process

- Obtain quotes for the work to be done

- Manage consultants who may be required during the architectural process, such as surveyors and engineers

- Manage the construction contract with the builders and tradespeople

- Ensure that you get the quality of work and standard of finishes that you expect

Turn to page 145 and complete the to-do items for Step 4 of the Home Build in a Box system.

CHAPTER 7

Step 5 – Builder

Having a home built today is not only about the finished product. It's also about the experience, or the journey, as I like to call it – the warmth of your first contact with the professionals who guide you on your journey, the beautifully presented display homes, the inspiring and thought-provoking floorplans. It's the whole package. The home building industry has evolved from being available to a select few who were willing to go the distance and put in the time, effort and money, to becoming a systemised and straightforward consumer experience available to everyday Australians. Not only can everyone now benefit from the advantages of building their home, but they can do it simply and easily. The builders I choose to work with provide a strong sense of trust and reliability. To help you have the same experience with your builder, this chapter shares the insights we've gained over the years of selecting and working closely with builders.

Custom builder or volume builder?

Volume builders are usually, but not always, nationally owned companies that have completely systemised their entire home building process. In addition to systemising the process, they utilise modern sales and marketing strategies to help families build new homes. The main feature that attracts people to a volume builder is their extensive range of floorplans that are already priced out to cater to budgets ranging from standard to luxury. Volume builders work on a standard inclusions-plus-upgrade scale where you can add and subtract according to your budget. It's a very simple option for people who want someone to take care of the whole process from start to finish.

Volume builders often partner with land owners to market house and land packages.

One thing to keep in mind when using a volume builder is that everything is finalised before you start. Every single detail is set out right down to the colour of the paint on your walls. If you are someone who likes to improvise and change things as you go, you might be better suited to a custom builder.

Custom builders are usually hands-on local tradesmen who pride themselves on the personal quality and uniqueness that they can bring to your home building experience. Custom builders suit those of you who like the idea of having some level of involvement in the home building process, though that is completely optional.

One of the greatest advantages of using a custom builder is having the ability to improvise and make adjustments along the way. While it's common for custom builders to have a small range of plans to inspire you, chances are you will have already engaged the services of an architect and/or building designer. If you have an existing relationship with your custom builder, they may work alongside you during the design phase of planning to build your new home.

Budget builder or luxury home specialist?

Another decision to make is whether to use a builder whose experience is with budget homes or more high-end constructions. This raises an interesting point. The fact is that the quality of your home is determined by your budget more than your builder.

Every home with a building permit must be built to a standard that ensures its structural stability. Yours will have to be built according to a strict set of standards, codes, guidelines and regulations, regardless of the quality of the materials, fixtures and fittings. Every home build has a dedicated building surveyor. A building surveyor protects your home during the building process. Each stage must be signed off before work can continue. If shortcuts have been made, or if the work doesn't meet the standard set by the permit, the builder can't continue.

What determines the physical quality of your home beyond the legal requirements of industry standards are the materials used and the quality of the fixtures and fittings. These things are determined by your budget. Every builder has provisions to offer you either quality or budget products. Submitting your Builder's Brief and budget at the beginning sets the scene for whether you'll be using 'quality' or 'budget' products to build your home.

Can a quality builder build a cheap home? Absolutely. Can a builder who specialises in affordable housing build an upmarket home? You bet. If a builder has a reputation for building 'cheap' homes the chances are that they have simply niched the affordable housing market, in much the same way that a renowned quality builder has probably niched the luxury housing market.

In saying that, if you have a generous budget and you are looking for a luxury build, it makes sense to engage a builder who has a niche in this market. Or, if you are a family who needs space over fancy features, you'll be better off contacting a builder who specialises in building homes on a budget. A builder's marketing strategy will often give you the best idea of what that builder specialises in. If you are on a tight budget, don't spend your time engaging builders who market architecturally designed homes filled with luxury features. Instead, choose builders who focus on value.

Selecting your builder

The first thing you need to know when engaging a builder is whether they are licensed. Only licensed builders can be in charge of significant building projects. While the laws do vary somewhat from state to state, a licensed builder has to provide a warranty for their work. You can check to see if your builder is licensed by visiting the building commission and practitioner boards listed in the further resources section.

Beyond the legal requirements, the only sure way to know if a particular builder is right for you is, of course, to build a home with them. However, the chances are you haven't done that, in which case you should take a recommendation from someone who has. If you don't have any recommendations, then you might like to do some homework. If choosing a volume builder, visit their display homes. Keep in mind, though, that display homes are often showcased with expensive upgrades and the same floorplan might look dramatically different on a tighter budget. The same applies if you are looking for the perfect custom builder. Ask if you can visit a few homes that they have built for others. At the very least, get some addresses and do some drive-bys to see what their work looks like.

Now let's look at the nitty gritty of the selection process.

By the time you are ready to meet your potential builders you will have:

1. Calculated your budget

2. Defined your needs, wants and desires

3. Know where you want to build

4. If using a custom builder, have your plans

Finding the right builder is now a matter of taking this information to your potential candidates and asking either the custom builders to provide you with quotes or the volume builders to provide you with floorplans and prices. Remember that custom builders will be quoting your job from your supplied floorplan and supporting information, while a volume builder will be using all your information to supply you with a range of floorplans that can be built within your budget. Volume builders will deliver options in a very short time. Possibly even instantaneously. Custom builders will take more time but will ultimately achieve the same result. Getting the right floorplan that accommodates all your needs and then manages your wants and desires via upgrades is a finely choreographed dance that your builder or new-home building sales consultant is trained very well in. Recently, I walked into the office of one of Australia's leading builders. Because I'd followed the Home Build in a Box system, within two hours I had a solid idea of what I was going to build and had chosen my scope of upgrades. Thanks to my groundwork, it was a very straightforward process.

Now let's assume that in this case you go through this process with a maximum of three volume builders. Any more than three and the decision-making becomes harder. And let's say you receive a maximum of three options from each builder. Each option includes a suitable floorplan that can be built within your budget, but may also include any upgrades and site cost additions. This will give you nine full options to choose from. Because you've been very clear with them on your wants, needs and desires, along with having a strict budget, there's no room for any grey areas. Of the options presented, you'll choose the option you love the most and that could be made up not only of the physical attributes of the home, but also the quality of customer service you receive.

In PART 3 on pages 149-150 you'll find a simple template that you could use to contact builders via email. It might be a good idea to call them first to let them know it's coming. You can use it as it is or improvise to suit your needs and personality.

Avoiding budget blow-out

There is a common misconception that circulates about builders giving you a price and then, before you know it, the price has been ramped up and they've blow your budget. What actually happens is that you blow your own budget. It goes back to what we were saying earlier about resisting the temptation to upgrade. Here is the mistake people make. They see a house and land package advertised for a certain price. They see that it is within their budget and make enquires to have it built. The floorplan of the home and the location offers you a great lifestyle. What happens next in this sequence of events is what hurts the most. You visit the display home. You fall in love with the extras and upgrades. You explore all the lovely façades. Now, rather than focussing on what you *can* have and what you *can* afford, you feel deflated by the things that you can't.

I've often thought about how they could change this. In sales it's always a good idea to give your customers what they expect and then surprise them with just that little something more. In other words, promise less; deliver more. Unfortunately, many of the volume builders do it the other way around. They promote their product pumped up with the extras, but often the majority of their customers can't afford all the upgrades that their advertising flashes up in lights. This is the way it is in most industries. Everyone wants to show off their best. My advice here is to stick to your guns and your budget and be aware of these seductive sales techniques, so that you can avoid being sucked in to paying for things you really can't afford or are low on your list of priorities.

Remember what I told you about a good, functional floorplan that flows beautifully being the most important aspect of home design – the thing that lasts well beyond the day that the new tiles are scratched and the luxury carpet is stained. While flash façades and classy upgrades are going to give you an elevated sense of pride in your new home, they won't actually improve the function and liveability of your home, and therefore also won't affect the quality of lifestyle that your new home will deliver to you and your family.

Choose a builder who makes you smile

When it comes to choosing a builder I need one thing to happen before anything else can. They need to make me smile. Why is this so important above all else? Because *genuine* people make us smile and genuine people are the ones we can trust. When we feel someone's authenticity we feel that they will look after us. My experiences with building have been good, but it is inevitable that little things will arise from time to time that mean you have to ask some hard questions. It could be a small thing such as the fact that when the light shines on the wall in the afternoon you can see the difference in the coats of paint. Or it might be a big thing, such as, 'Um, sorry, but they are not the tiles I chose.' Ouch. These things can be as much or as little of a hassle as your builder lets them be. If the person who calls the shots – be it the new-homes consultant or the builder – gives you a sense of comfort and reassurance in the beginning, then if anything goes wrong they'll be more likely to do the right thing by you and take the necessary steps to keep you happy.

On the other hand, if you feel an air of arrogance about a builder when you first meet them, there's every chance that you'll receive an arrogant, unhelpful response when something goes wrong. Beware the ego. Ego can be damaging to you as the client. If you are not sure whether ego is getting in the way or not, consider the tone of your conversation with the builder. Does it feel like it's more about them than you? If so, you might reconsider working with them. You should reconsider working with anyone who is arrogant, no matter how good you've heard their work is. You certainly won't have much fun taking them to task if something isn't how you expected or hoped it would be.

Few people realise that once they engage their builder, there's really not much left to do. People fear the process of building because they know little about it. The truth is you don't need to know a lot about it. It is the process of *planning* to build a home that is the part you play, not the home build itself. I chose to share my experiences about *planning* to build a home, because that's where your input matters most. Once you've engaged your builder you might just find yourself sitting back sipping pina coladas as you watch the magic unfold before your eyes.

However, we have one last tip for you before you mix that drink.

Near enough is not good enough

Over years of chatting to families about their home building experience, we've come to realise that the things you'll wish you had done differently will be the things you never knew to look out for. Others had a feeling that something wasn't quite right, but they just didn't know what their alternatives were so they didn't address the issue.

One of the biggest mistakes made by people who are having a home built for them is not speaking up when they know something isn't quite right. It's those situations where you see something and you think 'Mm, should that be like that?' Or some people just didn't have the courage to stand up and say, 'Hey, that doesn't seem right' or 'That wasn't what we had planned.' They are often your commonsense things.

Let me tell you about a recent custom-home build of mine. One of the things I love about custom-home building is that you can improvise along the way. For example, on one occasion we realised that it would be better to have a single large impressive bathroom rather than a pokey toilet and pokey bathroom. It was a change for the better. Tick, great job! We went on to shuffle a few other areas around. We moved a wall a bit this way and a bit that way to get the balance spot on. It's the builder's job to make sure that, if you change one thing, another isn't affected and it all flows as it should. Well, on this occasion, one slipped by the keeper. We'd moved the back wall of the kitchen 500mm. However, the builder forgot to adjust the plan to also move the kitchen bench back 500mm and consequently the plumbers put the pipes in the wrong spot. It resulted in an over-sized kitchen and an under-sized thoroughfare.

The trouble was that it was only on the day the slab was poured that I realised that this had been overlooked. According to my builder, it was a bummer but nothing could be done now that slab had been poured. He assured me that it would be okay. Um, sorry, did he just try to tell me that it would be okay? Was he suggesting that I spend $250,000 on a house to have a kitchen bench in the wrong spot? Ah, nope, sorry, not on my watch. The tradies spent the next day cutting into the slab to move the pipes the required 500mm to make the space function as it always should have.

Would it have been okay? Probably. Would it have been perfect? No, and what's the point of doing something if you are going to settle for anything that's less than perfect. It's not like your home's being donated. You have the right to speak up.

Had this been my first build, or had I not had the courage to hold fast, I might have accepted the mistake and that area would have been a bugbear to me. I would have hated putting it on the market as anything less than perfect. I would have failed in my quest to create the best liveable space for a family that I could.

I have known people who will make an arrangement with their builder to put $3,000 from the final payment into a trust account, to be released to the builder upon the satisfactory inspection of the home eight weeks after hand over. It can take living in a home to realise that some things aren't as they should be. Common examples are paintwork, superficial cracks in joins from the home 'settling' and little jobs being left undone. It's a simple way to guarantee that your home is completed to your utmost satisfaction.

Like lots of things in life, knowledge is power and being empowered means being in control. When you are in control you have courage and so it goes. The purpose of this book is to give you that control. If something isn't right, speak up. Trust your instincts, because they'll never let you down.

Turn to page 147 and complete the to-do list for Step 5 of the Home Build in a Box system.

"

Tomorrow you'll wish that you had
started today.

Home Build in a Box
essential steps

CHAPTER 8
To-do lists

Budget

In our minds, our home is built according to our desires. However, in reality our home is built according to one thing: our budget.

The number one mistake made when building is not starting with this step. Start your home building journey off on the right foot by establishing your budget first. Your budget sets the foundation for all the decisions you'll need to make and is the driver throughout your journey. Knowing your budget from the start ensures a smooth ride.

Before taking the following steps, read
CHAPTER 3: STEP 1 – BUDGET on page 49.

STEP 1
Budget to-do list

☐ Research grants and concessions to discover what you are eligible for by visiting firsthome.gov.au.

☐ Choose your preferred type of lender (bank, broker or credit union).

☐ Apply for pre-approval with three lenders using the instructions below.

☐ Enquire about building using a construction loan.

☐ Calculate your budget using the following formula: savings + grants + bank loan = budget.

☐ If necessary, talk to family, friends and developers about creative pathways for financing your new home.

☐ Sign on the dotted line with your preferred lender.

To organise pre-approval for your home loan:

1. Contact three lenders and tell them that you are building a home and you'd like to make an appointment to speak with their home loan specialist. Get their direct email address so you can send them all your financial information prior to meeting with them.

2. Put together your information pack from the checklist below.

3. Email the information to the three lenders three or four days prior to meeting with them so that they have a chance to assess it before the meeting.

Checklist – what to submit for pre-approval
For employees:

☐ Three-month statement showing salary

☐ Last three payslips and employer letter detailing payment summary

If you're self-employed:

☐ Personal and business tax returns for the last two years

If you have rental income:

☐ Current lease agreement

☐ Tax returns showing ownership

☐ Three-month statement showing rent

If you have Government income:

☐ Letter from Centrelink showing benefits

☐ Account statements showing payments

Identification:

☐ Passport, licence or birth certificate

What else should I bring?

☐ Details of all other loans, savings and investments

☐ Details of monthly living expenses

Priorities

Discover your secret weapon to simplicity.

Building offers the best value as it enables you to spend your money on the areas of your home that will bring you the most happiness.

To make the process of getting quotes fast and efficient, use our template to easily create a simple Builder's Brief that will show your builder what matters most to you when it comes to creating your ideal lifestyle.

Budget + Builder's Brief (including Welcome Home Checklist) = accurate and efficient quoting.

Before taking these steps read
CHAPTER 4: STEP 2 – PRIORITIES on page 69.

STEP 2
Priorities to-do list

- [ ] Complete your Builder's Brief template

- [ ] Complete your Welcome Home Checklist

- [ ] Check everything twice and make sure you've thought about commonly overlooked areas.

Builder's Brief template

My home building budget is _____

My needs (essentials) – tick and add how many:

☐ Bedrooms ☐ Built-in-robes in bedrooms

☐ Living areas ☐ Walk-in-robe in master

☐ Garage spaces ☐ Storage

☐ Bathrooms ☐ Heating & cooling

NOTES _____

_____

_____

_____

_____

_____

_____

_____

My wants (hoping to get as many of these as I can) – number in order of preference:

☐ Extra bedroom

☐ Extra bathroom

☐ Extra living area

☐ Study nook or study room

☐ Alfresco/outdoor living area

☐ Walk-in-robes in other bedrooms

☐ Larger than average rooms

☐ Larger than average laundry

☐ Data & telecommunications

☐ Separate toilet

☐ Sustainable features

☐ Extra storage

☐ Walk-in pantry

☐ Ducted vacuuming

☐ Downlights

☐ Stone benchtops

☐ Feature façade

☐ Solar power

NOTES

My desires (would if I could) – number the list in order of preference:

☐ Butler's pantry   ☐ Smart home features

☐ Theatre room   ☐ Ceiling height increase

☐ Parents' retreat   ☐ Outdoor shower

☐ Bar/entertaining area

NOTES _____

_____

_____

_____

_____

_____

_____

_____

_____

_____

_____

Welcome Home Checklist (things to consider)

Living areas

☐ Proximity to bedrooms

NOTES

_____

_____

_____

_____

_____

Dining area

☐ Table area dimensions

NOTES

_____

_____

_____

_____

_____

Kitchen

- [ ] Location of kitchen for bringing in groceries from car
- [ ] Pot drawers
- [ ] Soft-close doors
- [ ] Above-bench cupboards
- [ ] Functionality of rubbish bin
- [ ] Power points on island bench
- [ ] Functionality of pantry
- [ ] Power points in or beside splash back
- [ ] Insinkerator
- [ ] Overhang bench depth
- [ ] Walk-in pantry or butler's pantry
- [ ] 900mm oven

NOTES

_____

_____

_____

_____

_____

_____

_____

_____

Bedrooms

☐ Block-out blinds

☐ Number and location of power points relative to bed

☐ Cabinetry and design of robes

☐ Opening/closing wardrobe doors or sliding

☐ Plastered or mirrored wardrobe doors

☐ Data ports

☐ Bed placement

☐ Cavity slider separating bedrooms from living for sound insulation

NOTES

_____

_____

_____

_____

_____

_____

_____

_____

_____

_____

Bathrooms & toilet

- [ ] Tiled or inserted shower base
- [ ] Recessed storage nook
  in shower
- [ ] Heated towel racks
- [ ] In-floor heating
- [ ] Heat lamps

- [ ] Double sink
- [ ] Towel storage
- [ ] Separate family toilet
- [ ] Window in toilet

NOTES

_____

_____

_____

_____

_____

_____

_____

_____

_____

Heating & cooling

☐ Central heating   ☐ Hydronic heating

☐ Ducted heating   ☐ In-slab floor heating

☐ Reverse cycle air conditioning

NOTES

_____

_____

_____

_____

_____

_____

_____

_____

_____

Storage

☐ Adequate sized laundry for today and as family grows

☐ Everyday storage for items such as games, school work, bags, etc

☐ Office or administration storage

☐ Rarely used storage for items such as camping gear, suitcases, Christmas items, etc

☐ Linen cupboard

NOTES

_____

_____

_____

_____

_____

_____

_____

_____

_____

Indoor electrical

☐ Location of TVs

☐ Telecommunications & data points

☐ Wall for charging handheld vacuums

☐ Location of family computer

☐ Downlights

☐ Dimmer lights

☐ Ceiling fans

☐ Heating & cooling

☐ Over kitchen bench feature lights

☐ Smart home features

NOTES

_____

_____

_____

_____

_____

_____

_____

_____

Outdoor electrical

☐ Coming-home light sensors

☐ Around-the-house sensors

☐ Home security

☐ Exterior power points

☐ Exterior lighting

☐ Location of heating/ cooling system

☐ Location of water tank

☐ Outdoor taps in the right locations

☐ Alfresco TV

NOTES

_____

_____

_____

_____

_____

_____

_____

_____

_____

_____

Basics

- ☐ Cavity sliders
- ☐ Deep freeze
- ☐ Ceiling heights
- ☐ Closed off, separate living areas

NOTES

_____

_____

_____

_____

_____

_____

_____

_____

_____

Land

Discover your perfect place to call home.

This step is about discovering your place to call home. It's important to get expert advice from the very beginning. When choosing the right block, work with a land specialist who understands the home building process. Buying land based on inaccurate advice could see you blowing your budget.

Before taking this step read
CHAPTER 5: STEP 3 – LAND on page 79.

STEP 3
Land to-do list

☐ Choose a legal conveyancer early so that you have them in place to manage the contract of sale when you are ready to buy.

☐ Make a list of your preferred areas or estates that interest you.

☐ Research each area and fill in the Estate Comparison Checklist below to ensure you are comparing apples with apples.

☐ Create a shortlist of potential blocks, and use the Land Buying Checklist on page 146 to assess them.

☐ If applicable, contact a civil contractor or a site levelling specialist to give you an estimate of site levelling costs.

☐ Put your favourite block on hold while you complete the remaining steps of the Home Build in a Box system. You can commit to purchasing a block once you are sure that your block and your house plan are compatible.

Estate Comparison Checklist

Estate										
Price per m^2										
Block size range										
Playground										
Size of open space										
Walking trails										
Bike paths										
River views										
Rural views										
Ocean views										
Distance to CBD										
Is it part of a government housing affordability program?										
Site cost estimates										
Covenants										
Easements										

Land Buying Checklist

☐ Are most of the homes owner-occupied?

☐ Is my block located in a family-friendly area?

☐ Are there or will there be any playgrounds, trails or public open spaces that will deliver value?

☐ Are there any covenants in place that might restrict my choices?

☐ Are there any easements that will affect where I want to place my home?

☐ If applicable, do I know what my site levelling costs will be?

☐ Have I read everything in Chapter 5 about buying land to ensure that I am fully informed and know what to look out for?

NOTES

_____

_____

_____

_____

_____

_____

_____

Plans

A great home starts with a great floorplan.

When planning to build a new home, designing or choosing the right floorplan is the most important step in your journey.

**Before taking these steps read
CHAPTER 6: STEP 4 – PLANS on page 91.**

STEP 4

Plans to-do list

☐ Decide if you are using existing building plans, a building designer or an architect.

☐ Make an appointment with your preferred design professionals or builders. Take your completed Builder's Brief, including the Welcome Home Checklist, to the appointment. Include your budget and let them know where you'd like to build.

☐ Choose the professional who best suits your requirements.

NOTES

Builder

Choosing the right builder makes for a happy home building experience.

**Before taking these steps, read
CHAPTER 7: STEP 5 – BUILDER on page 107.**

STEP 5
Builder to-do list

☐ Decide whether you are using a volume builder or custom builder, then follow the appropriate steps.

Using a volume builder:

1. Using the template on page 149, email your completed Builder's Brief, including the Welcome Home Checklist, to three builders. Include your budget and where you'd like to build. Request that they contact you to arrange a time to meet so that you can view suitable floorplans that are within your budget.

2. Compare the proposals you receive from each builder while also taking note of their customer service.

3. Choose the builder who can deliver your best outcome in regards to floorplan, conditions, service and price. Before making a final

decision meet with your preferred builder over coffee to discuss the proposals further. In other words, get to know them a little to be sure they are someone you'll love to work with.

Using a custom builder:

1. Work with your building designer or architect to finalise your perfect floorplan and create working drawings with complete specifications.

2. Once you have your plans finalised, deliver your plans and working drawings to three custom builders. Using the template on page 150 include a letter inviting them to quote your job and let them know that you are available to meet them.

3. If you have a time constraint, ask that they submit their quote within a reasonable timeframe.

4. Compare quotes received and choose the builder who can deliver your best outcome in regards to conditions, service and price.

5. Before making a final decision meet with your preferred builder over coffee to discuss their quote further. Get to know them a little to be sure they are someone you'll love to work with.

6. Decide who will build your home.

Letter/email for a volume builder

Hi my name is _____

I would like to build a home in _____ (estate or area) _____

My budget is _____

I am currently in the process of obtaining three quotes from builders and I would like to invite you to submit a proposal including suitable floorplans according to my home building budget of $_____ excluding land.

To give you an idea of my priorities I have attached my Builder's Brief and Welcome Home Checklist. If you could please contact me to arrange a time to meet to discuss your proposal, that would be great.

You can contact me by email on _____

Or you can call me on _____

Look forward to hearing from you at your earliest convenience.

Kind regards

_____

Letter/email for a custom builder (when you have working plans ready to build from)

Hi my name is _____

I would like to build a home in _____ (estate or area) _____

My home building budget is $_____ excluding land.

I am currently in the process of obtaining three quotes from custom builders and I would like to invite you to submit a quote.

Please find attached my plans, working drawings and specifications.

If you could please contact me to arrange a time to meet to discuss, that would be great.

Given my time restraints, could you please submit your quote within _____

You can contact me by email on _____

Or you can call me on _____

Look forward to hearing from you at your earliest convenience.

Kind regards

_____

NOTES

Conclusion

By now, I'm hopeful that you can see how building a new home is a great way to create the best lifestyle that your family can afford today. It makes financial sense, it makes practical sense and it's going to be better for your mental and emotional wellbeing in the long run. Even if it's not your dream home, a home that you build for yourself will be perfect for you in ways that a home designed for someone else could never be.

I also hope that by now I've exploded any fears that you may have had about the home building process. If you follow the five-steps in our Home Build in a Box system, you can be sure that your home building journey will be stress free and give you the best possible result with your available budget. You will have everything you need in your new home, and as much of what you want and desire as you can afford. What's more, you'll probably enjoy the home building experience so much that you'll look forward to the day when you have a bigger budget and can build a bigger and better home. But whatever you do, don't wait for that magic moment to build your magic home. Do the best that you can now.

And now I'm going to leave you with a story that shares why that is so important.

I grew up close to a family who had a wonderful little home in regional Victoria. It held the memories of the birth of each and every child. The carpet was well worn with the wear and tear of little people, a tapestry of love and loyalty that was sewn into the fabric of life. Even when I went there to visit I felt at home. It was a sanctuary for all who were fortunate enough to be blessed to be within its walls. They had four children and as their children grew, so did their friends and the congregation of joy that would find its way there weekend after weekend. For a kid, it was the kind of family you secretly wished you were part of, the kind that has bodies piled to the rafters cuddled up watching movies on a Saturday night without a worry in the world. The kind where everyone eats where they can find a spot and where there is a sign at the front door that says 'excuse the mess, we're making memories'. Upsizing to feel the freedom that comes with a bit of extra space was always on the agenda, but with four busy kids they just never found the time. So they continued on their journey of making beautiful memories until they had some more time to think about it.

Then one day, they did get around to building their bigger home – once the kids had become older, quieter, less busy, less… around. All the procrastination and excuses were valid no more and it was finally the right time. It was a home with big open living areas, the freedom to entertain and accommodate a large crowd and required virtually no maintenance. It seemed they finally had just what they needed, except that they no longer needed it. Sure, when all the kids and their families come home it's the majestic hub of love and laughter they'd always hoped for, but during the times in between there was a sense of missed opportunity.

This is an example of a family waiting for all the planets to align before taking the step to create their best lifestyle. But there'll never be a more perfect time to build a home than right now. We are the masters of our own destiny and time waits for no-one. So what are you waiting for? Find out exactly what you can afford and start the journey towards building your best lifestyle today.

You can gain instant access to the complete online version of the Home Build in a Box system at buildinoz.com.au.

Glossary of
building terms

Architect – Degree-qualified design professionals who can manage every aspect of home building, from preliminary sketches right through to complete project management of design and construction.

Blueprint – The design or plan of your new home. The footprint of the home.

Builder's Brief – A formatted brief that outlines your requirements and priorities to a builder.

Building code – The National Construction Code outlines the requirements needed for standards of safety, health, amenity and sustainability in the design and construction of new homes.

Building designer – A designer who uses the process of drafting to design technical drawings that include the full specifications of the components and elements required to be able to build a home.

Building permit – A permit that gives permission for the construction of new dwellings.

Building surveyor – Responsible for making sure that buildings are safe, energy efficient and livable. Building surveyors work with engineers, architects, designers and builders to ensure that buildings are designed and constructed to comply with building regulations.

Buying off-the-plan – Purchasing land or a house before it has been built based on the information supplied on the plan.

Certificate of Title – Shows the plan of the parcel of land and also identifies the legal owner.

Construction loan – A short-term loan to fund the finances required to build your home. Interest is paid only on the expenses incurred with each stage of the construction. A construction loan is usually switched to become a long-term home loan at the conclusion of construction.

Conveyancer – Specialises in the legalities of buying and selling property on behalf of the buyer and the seller.

Custom builder – A builder who builds homes from custom-designed floorplans.

Deposit – The portion of your home loan that you must deposit in the bank to secure a home loan; for example, to secure a $400,000 loan you may be required to pay a twenty per cent deposit of $80,000.

Fixtures and fittings – Anything in your home that can be changed after the house has been built, such as a bath, a tap, carpet and curtains.

Guarantor – A person who is willing to put their home up as security against your loan, and agrees to cover your loan repayments if you default.

Holding costs – The interest you pay on money borrowed to buy property you are not currently able to use for the purpose you intended.

Home Build in a Box – The Build in Oz methodology used to walk you through each step of planning to build a new home.

Home equity – Home equity is the difference between the monetary value of your home and the money borrowed against it.

Home loan – Money that you borrow to fund the full purchase or build of your home, in which principal and interest are paid back to the bank with the intention of paying down the loan in twenty-five to thirty years.

House and land package – A house packaged with a block of land to give you an idea of the overall cost of both the house and land together.

Interest – What you are charged by the lender for borrowing the principal.

Land Buying Checklist – A checklist that outlines key considerations for buying a block of land.

Land Comparison Chart – A chart that compares the features of vacant residential land in an area.

Land specialist – A property consultant who works directly with developers and builders to help clients source the perfect block of land. A land specialist's core product is land.

Lenders Mortgage Insurance – Typically, if your deposit amount is less than twenty per cent, you may be required to pay Lenders Mortgage Insurance, which protects the lender if you are unable to meet your mortgage repayments and the property has to be sold.

Liveability – Your home's capacity for enjoyable living.

New-homes consultant – The primary responsibility of a new homes consultant is to facilitate the 'off-the-plan' sale of a new home.

Planning permit – A legal document that gives you permission for use or development of a particular block of land.

Pre-sales – When land in a subdivision is sold before the titles have been issued, or off-the-plan.

Principal – The amount you borrow to buy/build your property.

Residential estate – A designated area of residential land lots.

Settlement – The monetary transaction that takes place between the buyer and the seller.

Stamp duty – A land transfer tax that is payable to the Government each time a property is transferred from one owner to another. Stamp duty is calculated on the value of the property at the time of transfer.

Subdivision – The process of dividing one large parcel of land into smaller parcels.

Sustainability – The effective use of natural resources and design to aid in the energy efficiency of your home and to minimise your carbon footprint.

Volume builder – A builder who builds a large volume of homes from their own range floorplans that can be duplicated.

Welcome Home Checklist – A checklist of items for consideration when planning to build your new home.

Working drawings – The complete and full technical drawings used by a builder to build a home.

Further resources

For information on building regulations, contracts and licensing, insurance requirements and disputes, including due diligence checklists, contact your relevant state authority:

Australian Capital Territory

- Access Canberra, www.accesscanberra.act.gov.au

New South Wales

- Department of Fair Trading, www.fairtrading.nsw.gov.au

Northern Territory

- Department of Infrastructure, Planning and Logistics, www.transport.nt.gov.au/lands-and-planning/building

- Building Practitioners Board, http://bpb.nt.gov.au/

Queensland

- Building Services Authority, www.qbcc.qld.gov.au

South Australia:

- Office of Consumer and Business Affairs, www.cbs.sa.gov.au

- South Australia Planning Portal, www.saplanningportal.sa.gov.au/

Tasmania

- Department of Justice, www.justice.tas.gov.au/building

Victoria

- Consumer Affairs Victoria, www.consumer.vic.gov.au
- Victorian Building Authority, www.vba.vic.gov.au

Western Australia

- Department of Commerce, www.docep.wa.gov.au

Local Councils should be contacted for information on approvals, permits and permissions.

For information relevant to your state on first home owners grants and concessions, stamp duty and buying a property:

- www.firsthome.gov.au

Australian Capital Territory

- www.revenue.act.gov.au

New South Wales

- www.osr.nsw.gov.au

Northern Territory

- www.treasury.nt.gov.au/

Queensland

- www.treasury.qld.gov.au

South Australia

- www.revenuesa.sa.gov.au

Victoria

- www.sro.vic.gov.au

Tasmania

- www.sro.tas.gov.au

Western Australia

- www.finance.wa.gov.au

The National Construction Code (NCC) provides the minimum necessary requirements for safety, health, amenity and sustainability in the design and construction of new buildings (and new building work in existing buildings) throughout Australia.

- www.abcb.gov.au

Your Home is a Government guide to building, buying or renovating a home. It shows how to create a comfortable environmentally sustainably home that is economical to run, healthier to live in and adaptable to your changing needs.

- www.yourhome.gov.au

For Government information about buying, selling or renting housing, property and land:

- www.australia.gov.au/information-and-services/family-and-community/housing-and-property

The Australian Investment and Securities Commission (ASIC) provides financial guidance you can trust.

- Australian Investment and Securities Commission (ASIC), www.moneysmart.gov.au

The Australian Institute of Architects

- www.architecture.com.au

Building Designer Association of Australia

- www.bdaa.com.au

Victorian Building Commission

- www.buildingcommission.com.au

New South Wales Building Professionals Board

- www.bpb.nsw.gov.au

Queensland Building and Construction Commission

- www.qbcc.qld.gov.au

Northern Territory Building Practitioners Board

- www.bpb.nt.gov.au

South Australia Development Assessment Commission

- www.dac.sa.gov.au

BUILDINOZ
build your best lifestyle

About Build In Oz

Build in Oz is an online platform that shares the benefits of building a new home with Australian families. Its purpose is to show families how choosing to build can help them live the best lifestyle that their finances, environment and personal circumstances can afford.

Build in Oz was founded by boutique property developer Natalie Stevens. Natalie was born into the civil construction industry and today works as an executive within her family business that has, over the past twenty years, provided more than 1,040 families with the perfect place to build their home.

With the issue of housing affordability affecting everyday Australians, families are desperate to find a solution to their problem of needing a better home, in a better location for a better price.

Build in Oz offers the end-to-end solution to this problem by delivering the *Home Build in a Box* system via a user friendly online learning portal. The online portal delivers a five-part process that navigates families step-by-step along the path of planning to build a new home. It shows them exactly where to start, where to turn and what to look out for along the way. *Home Build in a Box* makes it easy for families to have the perfect plan in place to ensure a happy and successful home building experience.

You can learn more about *Build in Oz* and get online access to the *Home Build in a Box* system by visiting buildinoz.com.au.

About Natalie Stevens

Natalie Stevens is the founder of *Build in Oz* and the *Home Build in a Box* System.

As a boutique property developer Natalie is best known for sharing the rarely known benefits of building with regional Australian families. She specialises in showing families how building can help them to live the very best lifestyle that their finances, environment and personal circumstances can afford. What's more, her *Home Build in a Box* system explains exactly how to achieve it.

Natalie was raised in the civil construction industry within a family whose business began in 1967 and who has over the past twenty years produced more than 1,040 residential blocks of land in regional Victoria. She works along-side her husband, within the same family business that today manages all elements of the property development supply chain, from the purchase of raw land right through to marketing and sales.

Natalie holds a degree in Architectural Ceramics and Education. She lives in the beautiful seaside city of Warrnambool which sits at the end of the Great Ocean Road along the Shipwreck Coast in South West Victoria. Today, Natalie is happily living her best lifestyle with her husband Sam, and their four children: Archie, George, Matilda and Jimmy.

You can contact Natalie by visiting nataliestevens.com.au

Buy1GIVE1
B1G1®

Impact – B1G1

When you purchased this book something wonderful happened. You helped combat illiteracy for an orphaned child in Nepal by providing them with five days of internet access. Education has been pivotal in my journey and I believe that every child deserves access to information that gives them the gift of knowledge. Making this donation is a simple, yet powerful example of putting this belief into action.

This has been made possible through my partnership with the Global Giving Initiative B1G1: Business for Good. I believe that every business has the power to change lives by giving back through its everyday business activities. So thank you; not only for purchasing this book, but for also making a huge difference in a child's life. Together, we've made an impact.

www.ingramcontent.com/pod-product-compliance
Lightning Source LLC
Chambersburg PA
CBHW042312210326
41598CB00042B/7372